Sustainability of bio-jetfuel in Malaysia

Sustainability of bio-jetfuel in Malaysia

Edited by
Jean-Marc Roda

Authors
Jean-Marc Roda, Maxime Goralski, Anthony Benoist, Anaphel Baptiste, Valentine Boudjema, Theodoros Galanos, Marine Georget, Jean-Eudes Hévin, Simon Lavergne, Frédéric Eychenne, Kan Ern Liew, Cyrille Schwob, Marcel Djama, Paridah Md Tahir

AMIC - AIRBUS - CIRAD - UPM - MIGHT - BIOTECHCORP
2015

Centre of International Cooperation in Agronomy Research for Development (CIRAD), 42 rue Scheffer, 75116 Paris France
www.cirad.fr

Printed by CIRAD, 2015
ISBN: 978-2-87614-706-5

Upper cover photo by Airbus Group
Lower cover photo by Claire Lanaud (CIRAD)

Editor: Roda, Jean-Marc
Authors: Roda, Jean-Marc; Goralski, Maxime; *et* al.

Sustainability of bio-jetfuel in Malaysia
By Jean-Marc Roda, Maxime Goralski, Anthony Benoist, Anaphel Baptiste, Valentine Boudjema, Theodoros Galanos, Marine Georget, Jean-Eudes Hévin, Simon Lavergne, Frédéric Eychenne, Kan Ern Liew, Cyrille Schwob, Marcel Djama, Paridah Md Tahir - CIRAD, 2015.

CABI Thesaurus: 1. biofuel 2. aviation 3. sustainable 4. economics 5. environment 6. ecology 7. industry 8. policy 9. Southeast Asia 10. Malaysia 11. Peninsular Malaysia I. Jean-Marc Roda II. Maxime Goralski III. Anthony Benoist IV. Anaphel Baptiste V. Valentine Boudjema VI. Theodoros Galanos, VII. Marine Georget VIII. Jean-Eudes Hévin IX. Simon Lavergne X. Frédéric Eychenne XI. Kan Ern Liew XII. Cyrille Schwob XIII. Marcel Djama XIV. Paridah Md Tahir Title. Sustainability of bio-jetfuel in Malaysia

ISBN 978-287614706-5
9 782876 147065

Contents

Figures

Figures

Tables

Abbreviations

ALCA - Atributional LCA
AIT - Asian Institute of Technology Bangkok
AMIC - Aerospace Malaysia Innovation Centre
ASTM - American Society for Testing and Materials
AVTUR - AViation TURbine Fuels
BtL - Biomass to Liquid
BPS - Biofuel Production System
CH_4 - Methane
CIRAD - Centre of International Cooperation in Agronomy Research for Development (France)
CLCA - Consequential LCA
CO_2 - carbon dioxide
CoE - Center of Excellence on Biomass Valorisation for Aviation
CSR - Central Sugar Refinery
DM - Dry Matter
DOS - Department of Statistics Malaysia
EFB - Empty Fruit Bunches
EU - European Union
FAO - Food and Agriculture Organisation (United Nations)
FAOSTAT - FAO Statistical Databases (United Nations)
FDPM - Forestry Department Peninsular Malaysia, see JPSM
FELCRA - Federal Land Consolidation and Rehabilitation Authority (Malaysia)
FELDA - Federal Land Development Authority (Malaysia)
FFB - Fresh Fruit Bunches
FOB - Free on Board (International Commercial Term)
FRIM - Forest Research Institute Malaysia
FT - Fisher-Tropsch
FU - Functional Unit
GHG - Greenhouse Gases
GIS - Geographic Information System
GWP - Global Warming Potential
HA - hectare
HEFA - Hydrogenated Ester and Fatty Acids
HLPE - High Level Panel of Experts on Food Security and Nutrition of the Committee on World Food Security (FAO)
IEA - International Energy Agency
ILCD - International Reference Life Cycle Data System

IPCC - Intergovernmental Panel on Climate Change
JET A - Jet Fuel with -40°C freezing point (mainly used in USA)
JET A-1 - Jet Fuel with -47°C freezing point (worldwide used)
JPSM - Jabatan Perhutanan Semenanjung Malaysia (Malaysian forestry administration)
KLIA - Kuala Lumpur International Airport
KPPK - Kementerian Perusahaan Perladangan dan Komoditi (Ministry of Plantation Industries and Commodities Malaysia)
LCA - Life Cycle Assessment
LCI - Life Cycle Inventory
LUC - Land Use Change
LULUC - Land Use and Land Use Change
MDF - Medium Density Fibre
MFIC - Malaysian Furniture Industry Council
MIGHT - Malaysian Industry-Government Group for High Technology
MMMA - Malaysian MDF Association
MPIC - Ministry of Plantation Industries and Commodities Malaysia...
MPMA - Malaysian Panel Products Manufactures'Association
MRB - Malaysian Rubber Board
MSM - Malayan Sugar Manufacturing Company
MSW - Municipal Solid Waste
MT - million tonnes
MTC - Malaysian Timber Council
MTIB - Malaysian Timber Industry Board
MWIA - Malaysian Wood Industries Association
MWMJC - Malaysian Wood Moulding and Joinery Council
N_2O - Nitrogen Oxide
NFA - National Forestry Act
OPF - Oil Palm Fronds
OPT - Oil Palm Trunk
PEKA - Persatuan Pengusaha Kayu-Kayan & Perabot Bumiputra (Association of Malaysian Bumiputra Timber & Furniture Entrepreneurs)
POME - Palm shells and Palm Oil Effluents
PPF - Palm pressed fibres
RED - E U Renewable Energy Directive
RISDA - Rubber industry Smallholders Development Authority
RM - Ringgit Malaysia
RPR - Residue to Product Ratio
RSPO - Roundtable on Sustainable Palm Oil
SETAC - Society of Environmental Toxicology and Chemistry
TEAM - Timber Exporters Association of Malaysia
UK - United Kingdom
UNEP - United Nations Environment Programme
UPM - Universiti Putra Malaysia
USD - US Dollar

Acknowledgements

This document is jointly produced by Aerospace Malaysia Innovation Centre (AMIC), Universiti Putra Malaysia (UPM), and Centre of International Cooperation in Agronomy Research for Development (CIRAD).

The work was financed by AIRBUS, AMIC, UPM, CIRAD.

The Centre of Excellence on Biomass Valorisation for Aviation was initiated by a consortium grouping consisting of AIRBUS, AMIC, Malaysia Industry-Government group for High Technology (MIGHT), UPM, and CIRAD. The consortium invited UPM to host and to manage the Centre of Excellence, and invited CIRAD to provide its scientific guidance. The Centre of Excellence on Biomass Valorisation for Aviation and the participating organisations extend their sincere thanks to the many organisations that provided their expertise to the group through meetings and other informal discussions.

Introduction

Aviation represents a small but growing share of global CO_2 emissions (2-3%), and South East Asia is where this industry grows the fastest. The industry targets 50% reduction in net CO_2 emission by 2050, and will need at least 2 million tonnes of biofuel by 2020. Commercial aviation is predicted to grow at a rate of 5% annually until 2030 and expect improvements in fuel efficiency by 1.5% per year till 2020 (targets set by the International Air Transport Association). The aviation industry needs to take continuous steps to maintain growth in an era of increasingly volatile oil prices and uncertainty in supply, as well as reducing its carbon footprint.

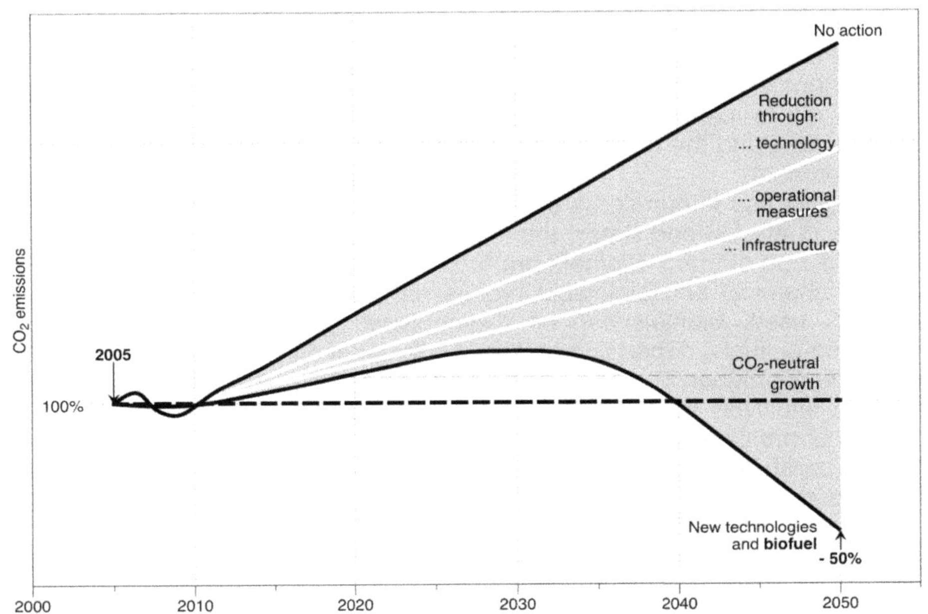

Figure 1: CO_2 emissions targets for aerospace sector.

For many years, this industry has been "criticised" for its high carbon Greenhouse Gas emissions. Of the total global emissions, 56% comes from burning of fossil fuels. The aviation industry alone contributes up to 649 million tonnes of greenhouse gases (GHG) emission annually, which represents 2 to 3% of the global CO_2 emissions. By 2050, the industry targets to reduce by 50% its net CO_2 emissions (compared to 2005 baseline).

The South China Sea divides Malaysia into two geographical entities; Peninsular Malaysia, or West Malaysia, and territories of Borneo (Sabah and Sarawak), or East Malaysia. Malaysia, located at the heart of South-East Asia, hosts one of the top three South Asian hubs for air transports (Kuala Lumpur) and is immediately neighboured by the two others (Singapore and Bangkok). The country is blessed with abundant biomass resources, which could be converted into alternative energy or other bio-products. However, even though government policies and market incentives have been put in place to support the use of green technology in the industry, the uptake of biomass commercialisation needs further intervention. Annually, a minimum of 168 million tonnes of biomass waste is generated in Malaysia. In Peninsular Malaysia, forest covers approximately 45% (5.8 million ha) of the land, while another 35% (4.5 million ha) is agricultural land. Out of the 4.5 million ha of agricultural land, oil palm (62%) and rubber plantations (29%) cover the largest area. Rice, sugarcane, or coconut form the bulk of the remaining area, and may generate proportionally higher tonnage of biomass residues than expected. However the exact quantities of sustainable biomass and residues and their availability for biofuel conversion is difficult to assess. This potential resource could represent a great opportunity for harnessing biomass energy in an eco-friendly and commercially viable manner. The first step to achieve such ambition is to identify the most suitable feedstocks to produce bio-jetfuels, and to understand the peculiarities of aviation bio-jetfuel sustainability in such an environment.

Biofuels give rise to numerous socio-economic and environmental issues. For example, in the US and Brazil, the impact of biofuels on food prices and food security is one of them. Biofuels are fuels derived from solid biomass through different chemical and biological processes and treatment according to the feedstock used. Biofuels can be distinguished as first-, second-, or third-generation biofuels, depending on the raw material and conversion technology used during production. Although second generation waste-sourced biofuels might be intuitively more sustainable than dedicated crop-sourced ones, the question of profitability, general acceptance of the pathways, and their impacts must be carefully weighted in order to confirm or invalidate their sustainability. Among potential impacts are: effects on greenhouse gases (GHG) emissions, atmospheric pollution, water consumption and pollution, deforestation, biodiversity loss, soils degradation, rural development issues, energy, security, health and social conflicts.

Would a single crop be sufficient source for biofuel conversion to fulfil the

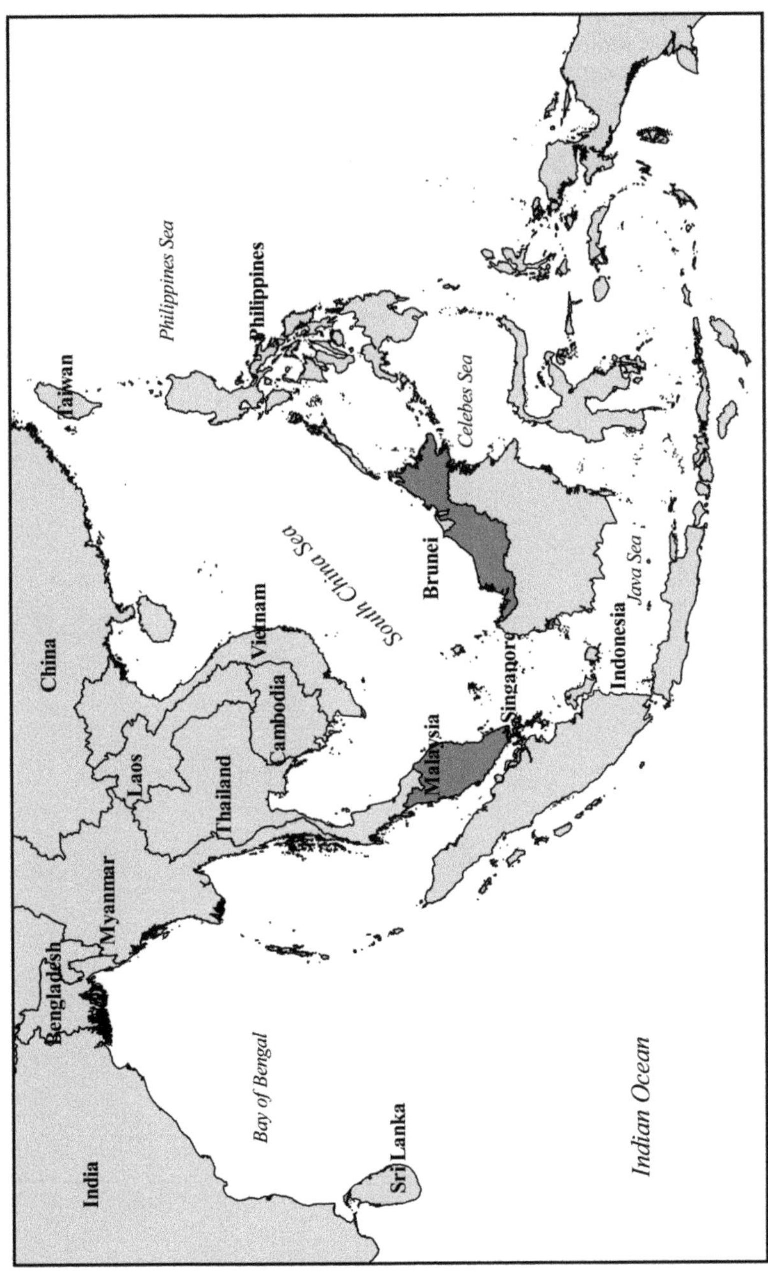

Figure 2: South China Sea divides Malaysia (in dark grey) into two geographical entities.

demand projected for 2030? If not, is multi-crop the solution? How would a multi-crop source influence on the conversion process and cost of production? Is the available biomass distributed in specific regions or is it scattered throughout the country? In this case, is the cost of collection and transportation affordable? If not, can multi-crop be the solution? Will this new bioenergy create land grabbing issues and environmental impacts, such as deforestation, thus creating socio-political issues? The present book answers some of these questions, states about the resource, its potential for bio-jetfuel and challenges. Detailed industrial simulation and prospective scenarios will be addressed in an additional volume (forthcoming).

1
From biomass to bio-jetfuel

On a global scale, biofuels create a lot of hope, but also a lot of concern

The United States and Brazil have led the biofuel sector since the 1970s, among other actions to cope with the oil crisis of 1973 and 1979. They encouraged first generation biofuel technology to transform corn and sugarcane, into bioethanol and biodiesel. Between 2005 and 2008, major biofuel programmes were initiated in the European Union and the United States. By 2012, around 60 countries had targets in place for energy security, improving the balance of payment, creating new sources of income and employment, developing rural areas and diminishing Greenhouse Gas emissions. However despite such noble aims, it also raises issues about competition with food production and arable lands. World food prices reached their peak in 2008 (Bailis et Baka, 2011). Grain and edible oil prices increased 70 to 120 percent; world food markets experienced the largest price shock in thirty years (Food and Agricultural Policy Research Institute, 2009). Several scientific assessments were published that confirmed doubts about the ability of biofuels to meet some or all of their stated objectives by demonstrating how, under existing and proposed production systems, biofuels could contribute to large-scale Land Use Change (LUC) (Fargione et al., 2008). It can happen through direct land use change (DLUC) or indirect land use change (ILUC) in which biofuel crops affect market conditions either by displacing crops or livestock or by diverting existing crops from one market (e.g., food or feedstock) into biofuel production (Bailis et Baka, 2011).

In response to these negative concerns, second generation biofuels emerged in order to mitigate the negative effects of first generation biofuels.

First generation biofuels refer to the fuels that have been derived from crop sources like starch, sugar or vegetable oil. The oil is obtained using the conventional techniques of production (from crops). Some of the most popular types of first generation biofuels are biodiesel, vegetable oil, biogas, bio alcohols and syngas (biofuel.org, 2013). They convert annually millions of

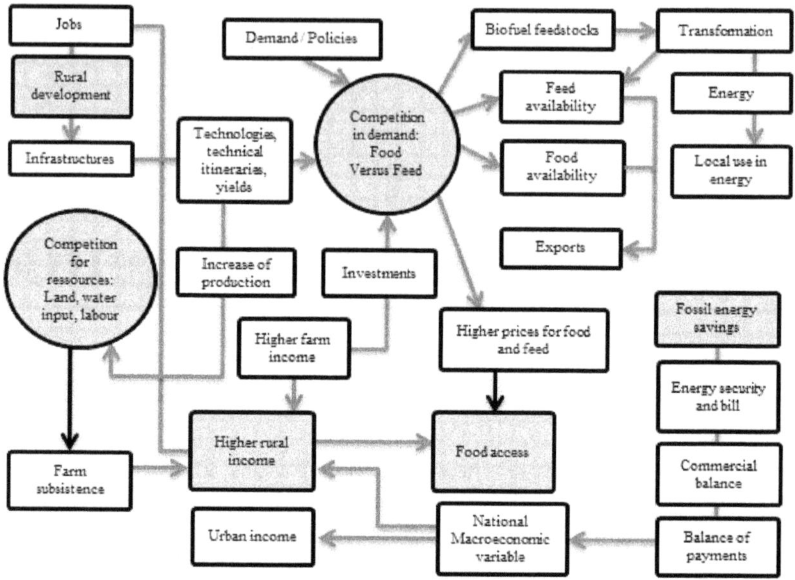

Figure 3: Main impacts and feedback in the food, agriculture and energy systems following the introduction of a biofuel demand (adapted from HLPE, 2013).

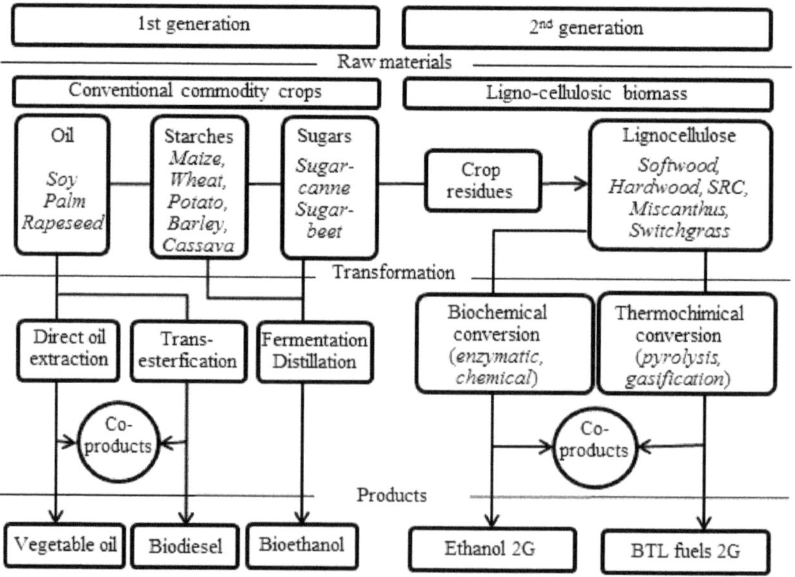

Figure 4: Pathways for producing first and second generation biofuels (adapted from Naik et al., 2010).

tonnes of vegetable oils, tallow, grains and sugar cane to biofuels, being 99.85 percent of the biofuels produced and consumed worldwide in 2011 (91 300 000 tonnes/year). In the same year, second-generation biofuels from lignocellulosic raw materials reached only 137 000 tonnes/year (IEA, 2013).
Conversely, second generation biofuels appear having less adverse impacts. Mostly produced from agriculture residues, forestry, industrial wastes or non-food energy crops. Being cheaper and presumed more abundant, they are supposed not to compete with food and to allow reduction in Greenhouse Gas Emissions (even if it still has to be proven through Life Cycle Analysis, see Bailis et Baka, 2011). The raw materials for the second generation include cellulosic materials, switch grass, waste biomass, wheat stalks, corn stalks, wood, and special energy or biomass crops such as Jatropha. Second generation fuels have their own drawbacks. Firstly, specialised biomass crops have already shown their limits. And secondly, to move forward to second-generation biofuels, is not so easy for developing or transition countries, given the often proprietary nature of this technology, the elevated capital investments required, and the high demands that second-generation technologies make on infrastructure, logistics and human capital (HLPE, 2013).

Bio-jetfuel specifications

Humans already use 12 to 20% of the terrestrial Net Primary Production (Haberl, 2007), which consist almost exclusively of plants. Terrestrial plants essentially produce sugars, starch, and oils, while their main constituent is lignocellulose and water. Only 10% of the harvested biomass from the Net Primary Production is already used as energy, while the remaining is used for materials (20%), animal feed (58%), and human food (12%) (Krausmann, 2008). Consequently, any additional uptake of terrestrial biomass for bioenergy use, may increase the pressure on ecosystems if it means producing and harvesting more sugar, starch or oil than it is already the case. The indirect adverse effects constitute a non-exhaustive list of the sustainability requirement for bio-jetfuels.

Sustainability criteria	Policy criteria
Do biofuels threaten food security?	Protecting the poor and food-insecure
Can biofuels help promote agricultural development?	Taking advantage of opportunities for agricultural and rural development
Can biofuels help reduce greenhouse gas emissions?	Ensuring environmental sustainability Enhancing international system support to sustainable biofuel development
Do biofuels threaten land, water and biodiversity?	Ensuring environmental sustainability Enhancing national system support to sustainable biofuel development
Can biofuels help achieve energy security?	Reviewing existing biofuel policies

Table 1: Bio-jetfuel sustainabilty requirements - Source: (Anonymous, 2012).

First generation fuels use oil, starch and sugars, and could easily become unsustainable if produced in large quantities, because of the potential stress that world scale production would place on food commodities (Gomez, 2008). Sustainability criteria for millions of tonnes of aviation fuels would require that they are not made from first generation fuels. Second-generation biofuels, produced from non-food, cheap and abundant plant biomass are essentially based on lignocellulosic biomass when referring to terrestrial biomass, and on oils when referring to algae. These biofuels are seen as the most attractive solution to this problem, but a number of technical hurdles must be overcome before their potential is realized (Gomez, 2008). For now, there is a general consensus in the aviation industry that any biofuel used in the sector should comply with sustainability criteria developed for road transport available for the US and the EU.

Sustainability criteria are a first step towards sustainable biofuel production, however technical options may be limited because of the high quality fuels required in aviation. To be acceptable to Civil Aviation Authorities, aviation turbine and jet fuels must meet strict chemical and physical criteria. There are basically two type of conventional jet fuels used in commercial aviation: Jet-A used mainly in the USA and Jet A-1 used worldwide; the only difference between them is the freezing point which is -40°C for Jet A and -47°C for Jet A-1. For the application of a fuel within the international civil aviation sector the certification in accordance to the ASTM International standards or the UK Defence Standardisation.

Requirement	Rationale	Specification
Energy content	Affects aircraft range	Minimum energy density
Freeze point	Impacts upon ability to pump fuel at low temperature	Maximum allowable freeze point
Thermal stability	Coke and gum deposits can clog or foul fuel system	Maximum allowable deposits in standardized heating test
Viscosity	Impacts ability of fuel nozzles to spray fuel and of engine to relight at altitude	Maximum allowable viscosity
Combustion	Creation of particles in combustor and in exhaust	Maximum allowable sulphur and aromatics content
Lubricity	Impacts upon ability of fuel to lubricate fuel system and engine controls	Maximum allowable amount of wear in standardized test
Compatibility	Fuel comes in contact with large range of metals, polymers and elastomers	Maximum acidity, maximum mercaptan concentration, minimum aromatic concentration
Safety	To avoid explosion in fuel handling and tanks	Minimum fuel electrical conductivity and minimum allowable flash point

Table 2: Bio-jetfuel technological requirements - Source: (Rosillo-Calle et al., 2012).

Converting biomass to bio-jetfuel

Different ways of conversion

Several pathways exist to convert biomass into bio-jetfuel. Within this variety of processes, few have been recognised and certified for the manufacture of aviation turbine and bio-jetfuel that involve blending conventional (fossil-based) and other synthetic components. Approved bio-jetfuel can be blended as high as 50% with traditional jetfuel. Three pathways are currently approved by the internationally accepted standards (ASTM) for alternative jet-fuel:

- Biomass-to-liquid (BtL), via Fischer-Tropsch process (FT-SPK) - approved in 2009.
- Hydro-processed esters and fatty acids (HEFA, also called HRJ or Bio-SPK) - approved in 2011.
- Synthesised Iso-Paraffinic fuel (SIP fuels or DSHC pathway (Direct Sugars to HydroCarbons) - approved in 2014.

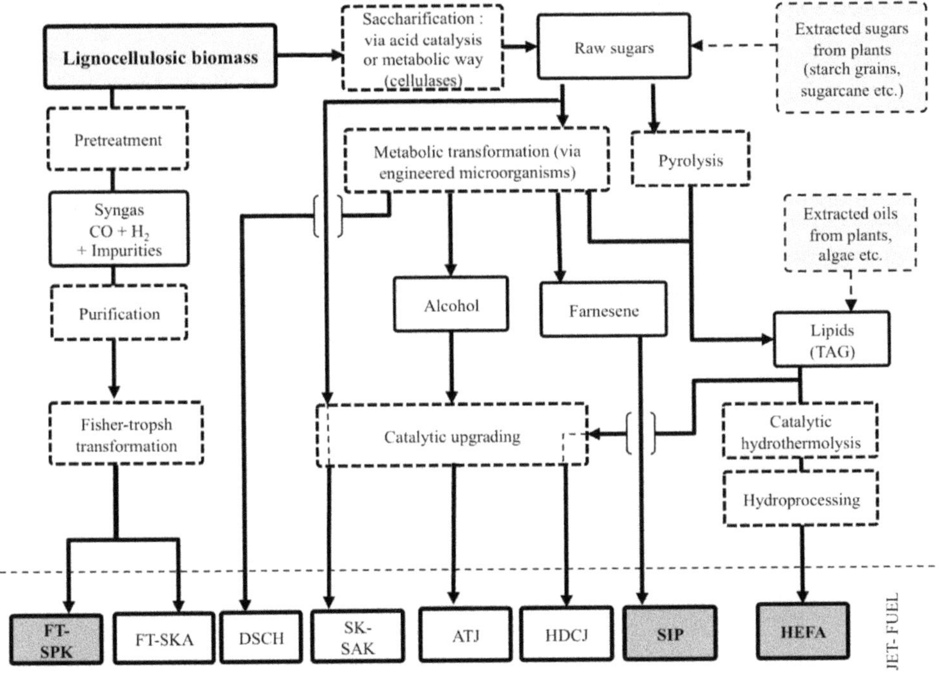

Figure 5: Existing certified and other possible pathways for AViation TURbine Fuels (AVTUR).

The Fischer-Tropsch process for bio-jetfuel involves converting any kind of lignocellulosic biomass through a thermal process and an additional step of gasification. It is a simple and reliable process, but its major inconvenience is that it requires a transformation to several physical phases (solid to gas to liquid), thus losing a lot of energy and material in the HEFA process can use any form of vegetable oil or animal fats. In the first step of production, the oils and fats are hydrogenated. They are refined in a second step in a similar process used with fossil fuels. The process is essentially liquid, thus very efficient but is currently handicapped by the very limited availability of cheap feedstocks. This process is currently not easy to set up for lignocellulosic biomass, but could be very promising for algae biomass, when large-scale solutions will exist to produce such biomass.

The DSHC process relies on the fact that sugars can be converted via metabolic pathways into farnesene. Farnesene ($C_{15}H_{24}$), is a precursor for jetfuel. This process is potentially the most promising, and is perfectly suited to any biomass composed of simple or complex sugars, such as lignocellulosic biomass.

Limited expansion potential for Southeast Asia croplands

Within the intertropical zone, with the monsoons ensuring 2 to 4 meters of rainfall per year, Southeast Asia harbours extremely high productivity of biomass, in terms of tonnes per hectare and per year. Consequently, forest and agricultural sectors have historically been preponderant in the economy and the land use of the region. However human pressure in South East Asia is also extremely high. Agrosystems form a mosaic of different crops, and are extremely fragmented in comparison to the vast and more homogeneous areas observed under similar latitudes in Latin America. As a result of this combination of factors, South East Asia, with the rest of developing Asia, is the

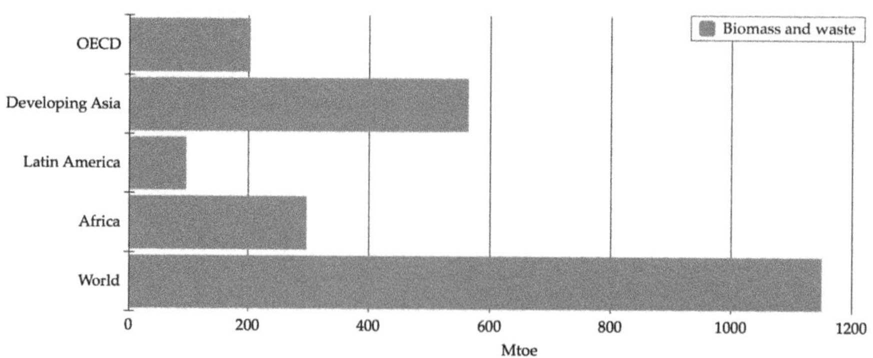

Figure 6: Total primary energy demand for biomass and wastes, by region - Source (IEA 2007).

region in the world where the total primary energy demand is the highest in the world for biomass and wastes (IEA 2007). The large development of agriculture in the last decades has already spread over the most suitable soils. Conversely, sustainable practices aiming to preserve the last remains of one of the most biodiverse and ancient rainforest in the world, will prevent local societies to deforest and to convert vast areas of additional lands to crop production. As a result, East Asia, South Asia and transition countries are the regions with less potential for cropland expansion in the world, except for the case of Near East and North Africa (FAO 2003). To date, the large-scale industrial development of algae production for fatty biomass is not yet a reality. In this specific South East Asian context, it becomes difficult to consider strategies entirely based on raw material coming from energy crops, be it primary production of sugars, starches or oils. Conversely this call for investigating in details the feasibility of pathways based on agricultural and agro-industrial residues, which are not, or not much used. The overwhelming tonnages of potential residues consist mainly lignocellulosic residues, which after all, are nothing very different from a mix of polymerised sugars.

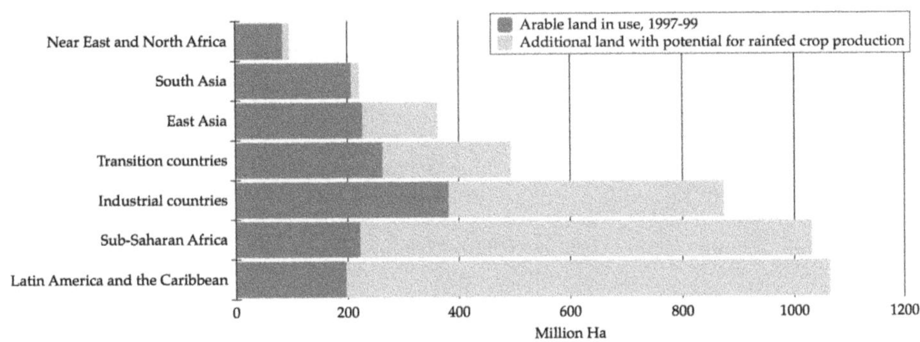

Figure 7: Potential for rainfed cropland expansion, by region - Source (FAO 2003).

What is lignocellulosic biomass?

Cellulose is the most abundant organic polymer among terrestrial biomass. Cellulose and lignin provide structural rigidity for every plant from the blade of grass to the giant redwood, which can reach 90 meters in height. They constitute more than 70% of the world's terrestrial biomass. Cellulose microfibrils, hemicellulose chains, lignin and to a lesser extent, pectin, are the major components of cell walls. There is a huge variety of lignocellulosic materials such as wood or agro-industrial residues like straws, shells, husks, sugarcane bagasse, *etc.* Despite a relative similarity in their biochemistry, these varieties of materials differ by their chemical and physical characteristics that can have a great influence on conversion processes (*e.g.*, chemical and biochemical composition, ash and extractives rates, moisture content, heating value). Besides, some other anatomic characteristics do not interfere with thermochemical reactions but play a role on a physical point of view (density). Cellulose is a polymer composed of 100 to 10,000 linked units of C_6 sugars (hexoses: glucose, mannose, galactose,*etc*). In their native state, cellulose molecules form fibres largely composed of compact crystalline domains separated by amorphous regions. Hydrogen bonding between cellulose layers accounts for crystalline cellulose's resistance to degradation. In fact the fundamental difference between starch and cellulose relies on the type of links binding the glucose units. Hemicellulose is a polymer with more than one type of subunit, predominantly C_5 sugars (pentoses: xylose, arabinose) and a smaller amount of C_6 sugars (glucose, galactose, mannose). Lignin is an aromatic polymer (polyphenolic) with high molecular weight and calorific (heat-release) value (Vancov, 2008). Lignin plays a role of cement within the plant cell wall. Moreover, it provides waterproofing properties and high resistance to biological degradation.

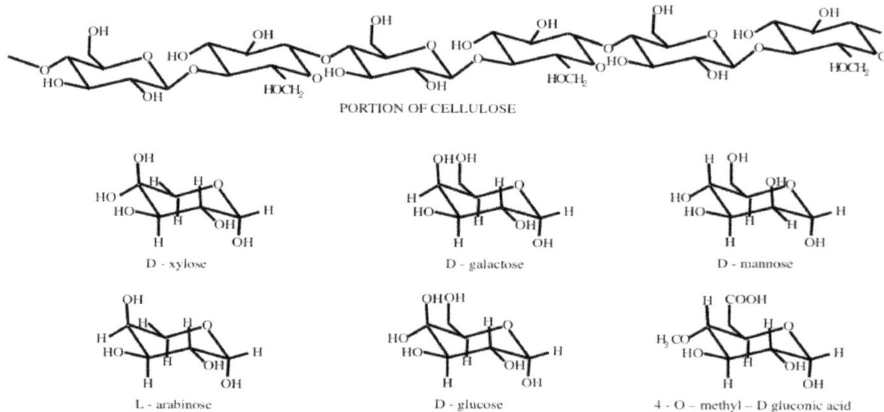

Figure 8: Chemical structures of cellulose and sugar components of hemicellulose - (Adapted from Mousdale, 2008).

2
Potential feedstocks and non-aviation biofuels

Biomass sources

Agriculture

Despite pedoclimatic conditions that would allow a large diversity of crops, Malaysian agriculture is nowadays focused on a restricted number of key crops: oil palm, paddy, rubber, sugarcane and coconut. The five major crops represent more than 82% of cultivated lands and concentrate almost 99% of gross output of the whole crop and horticulture sub-sector, on 24% of the Malaysian territory. Agriculture accounts for only 10% of the GDP. Oil palm forms the overwhelming part of the agriculture revenue, not counting the forest sector. During past decades, oil palm progressively replaced rubber in the hierarchy of cash crops, following the global changes in demand for this cheap oil, due to the increase in demand for this edible oil powered by the demographic dynamics of Asia and Africa. The Malaysian historical success story in eradicating the hardcore poverty in its rural communities essentially came from the increase in area of palm oil smallholders' schemes, at the expense of rubber plantations and forest areas. The average smallholder farm size is about 1.45 hectares. About 1 million small farmers cultivate 75% of the total area under agriculture. They represent the main contribution to food crop production as well as cash-crop production. The high geographical fragmentation of their agricultural activities has an impact on the supply and the valorisation of low price agricultural wastes, despite the existence of an excellent road network in Peninsular Malaysia.

Agricultural Products	Malaysia production (t)	World production (t)	Distribution
Palm Oil	16,993,715	45,768,605	37.13%
Rice	2,464,830	701,998,667	0.35%
Natural Rubber	939,2	10,286,913	9.13%
Sugarcane	800	1,700,648,436	0.05%
Coconut	550,14	60,295,788	0.91%
Other crops	2,137,885	NA	NA

Table 3: Agricultural production in Malaysia (data from FAOSTAT, 2014).

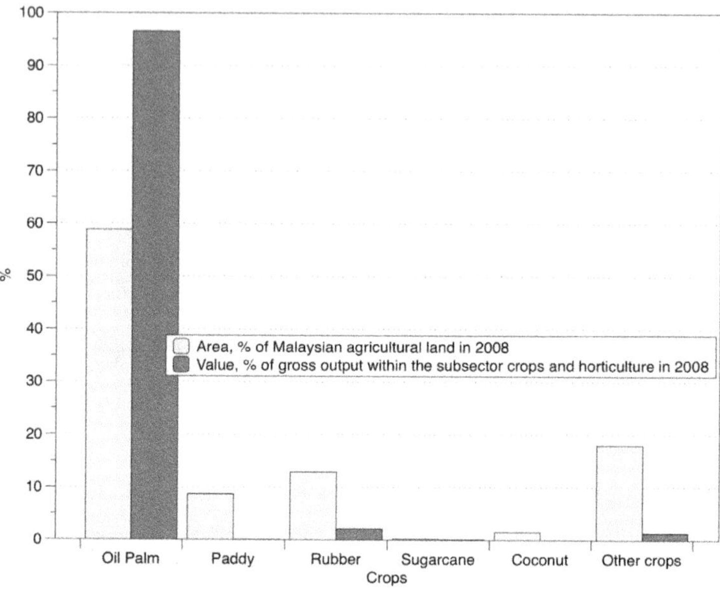

Figure 9: Areas and growth output in share of the horticulture and crop sector.

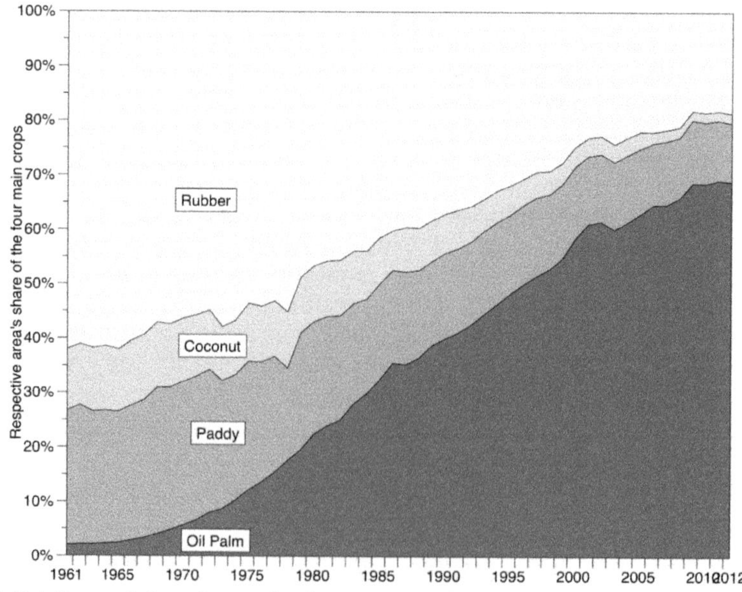

Figure 10: Relative evolution of areas for the top 4 crops in Malaysia.

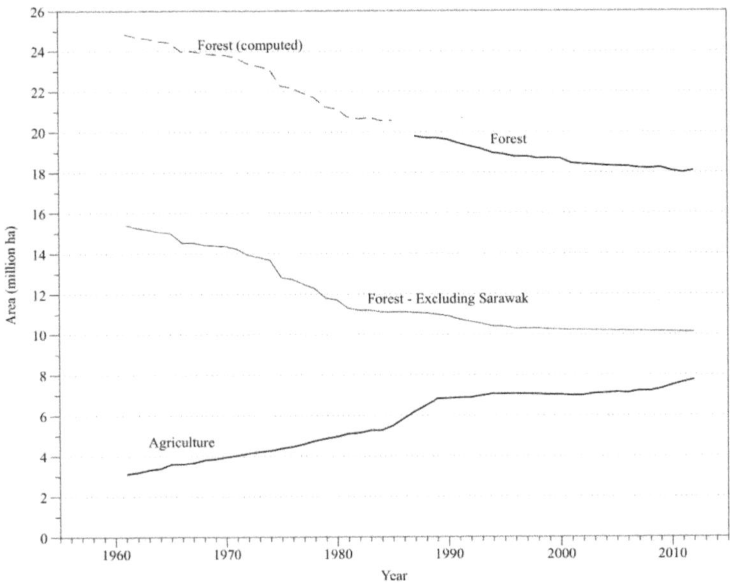

Figure 11: Forests conversion to agriculture since 1961.

Forests

Forest covers 62% of the total Malaysian territory.(FAOSTAT). Forest conversion to agriculture and to urban areas essentially occurred in Peninsular Malaysia during the 20th century, from 80% of the peninsula under forest in 1935, to less than 50% in the eighties, and has since stabilised to 44%. Forest conversion has become statistically insignificant in Peninsular Malaysia but continues in Sabah and Sarawak, where higher forest proportions of respectively 59% and 69%, higher poverty rates, and lesser economic development, encourage policies of cash-crops expansion (Malaysia Sustainable Forest Management, MTIB 2007). The contrasted situations between Peninsular Malaysia and Sabah-Sarawak result from deliberate state policies with different management frame (MTIB statistics for 2013). These facts suggest that any bio-jetfuel supply based on wood residues should focus on the peninsula, to avoid accusations of encroachment on the rainforest. The forest in the peninsula covers 5,83 million hectares (Forestry Departement Peninsular Malaysia statistics for 2013), which are all under sustainable management.

- Permanent Reserved Forest (PRF): 4.94 Mha,
- National parks & wildlife reserves: 0.59 Mha,
- State lands (or alienated forests): 0.3 Mha.

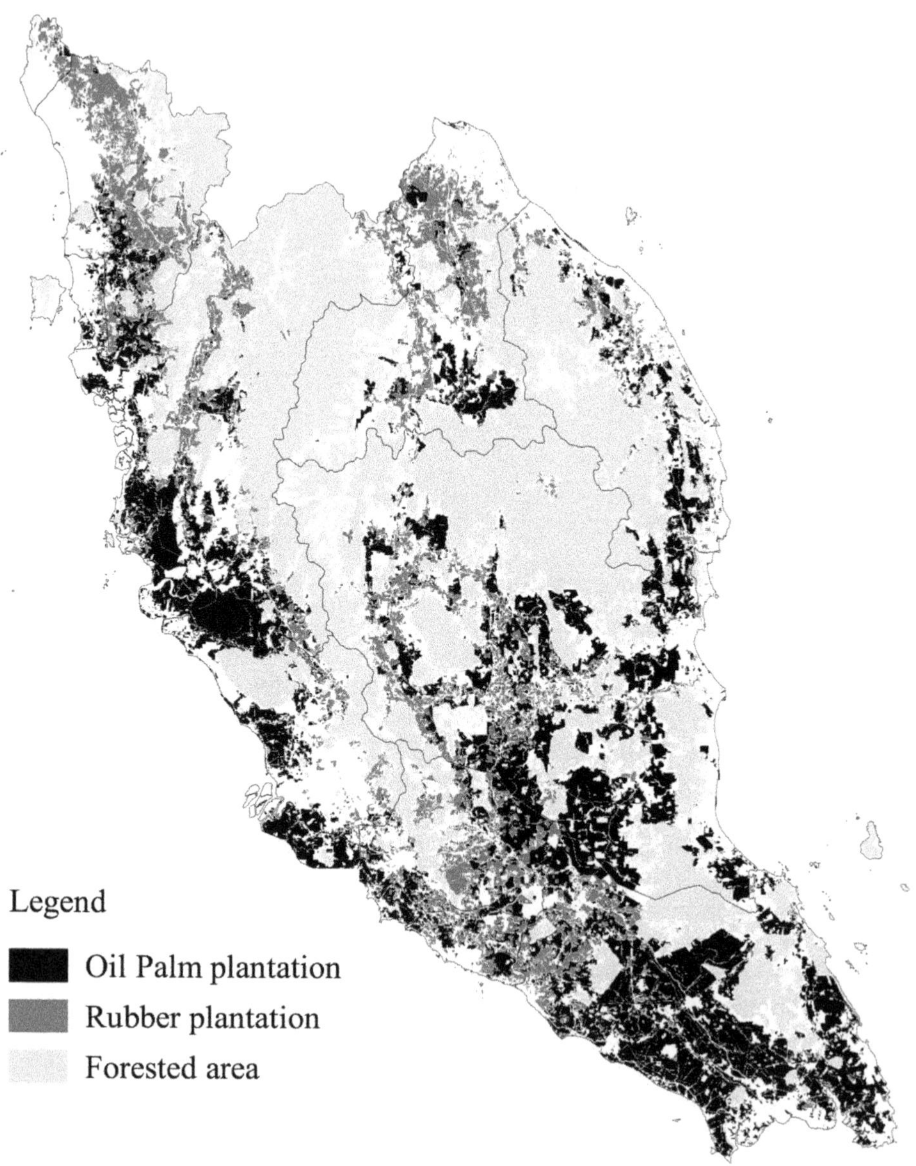

Figure 12: Map of forest, palm oil and rubber lands in Peninsular Malaysia.

Forty percent of Permanent Reserved Forest is dedicated to protection. The rest being subject to PEFC certified sustainable management and production. State lands forests are available for conversion into new uses (5% of the total forested area in peninsula). In Sabah and Sarawak, those proportions may be different.

Developments in non-aviation biofuels

Indonesia and Malaysia account for more than 80% of total palm oil production in the world. (Mukherjee et Sovacool, 2014). Malaysia is the biggest exporter of palm oil in the world. The two major products being derived from the palm oil industry are crude palm oil and palm kernel oil. The growing global market for biofuels and especially agrofuels created opportunities in Southeast Asian countries that find themselves as major producers, consumers and exporters. Malaysia currently produces essentially first generation biofuels and export biodiesel mostly to the European Union where there are subsidies for this product. Since 2006, through the National Biofuel Policy, the Malaysian Government has introduced the use of B5 blended biodiesel (95 percent petroleum diesel and 5 percent biofuels) (MPIC, 2013). Between 2006 and 2007, 92 biodiesel projects have been approved in Malaysia but by 2012, most of them have closed down because of high feedstock prices, over-optimization and competition in demand for raw oil by the agro-food sector. To revive these projects, the Malaysian government announced in 2013 that B10 blended biodiesel would become mandatory to encourage again the biodiesel industry, but at the date of impression of this book, this regulation is not yet effective.

Altogether, only 30 biomass transformation plants are currently operating; most of them are located in the states of Johor and Selangor. The main products biodiesel and a few other products emerge at a smaller scale: solid biofuel, charcoal biofuel, jatropha biofuel, ethanol biofuel and rice biofuel. The somewhat mixed results has forced the Government to re-direct their thinking to develop multi-feedstock and second-generation biofuels (Sheng Goh et Teong Lee, 2010). Policies developed around first generation biofuel could be used as a basis for policies for the second generation.

Along this line, five companies (Bell Group, Golden Elate, Genting Berhad, Kelas Wira and Teck Guan Group) have signed a joint venture cluster agreement to accumulate around 1.5 million tonnes of dry oil palm biomass that will be used to produce biofuels. The group will be set up in Sabah, where the Agensi Inovasi Malaysia has identified 70 potential mills, out of over 120 mills, that could be partners for this biomass joint venture cluster. Many more companies are expected to follow suit into the venture that is estimated to contribute an additional RM15 billion Gross National Income to Sabah's economy by 2020. (Lane, 2013). Other companies announced that they are working on turning biomass into second-generation biofuels, These companies are:

• Platinum, located in Damansara (Kuala Lumpur), which wants to convert waste oils and fatty acids derived from the palm oil industry into second generation biofuels,

• Bionas, which announced plans to convert jatropha oil into biofuel,

• Sahabat, a joint venture between Premium Renewable Energy Sdn Bhd and FELDA to turn oil palm biomass into second-generation biofuels.

• Algaetech, which announced that they will work on third generation biofuels and bioproducts from algae.

3
Oil palm biomass

Oil palm

In 2012 oil palm land represented 2.5 Mha in Peninsular Malaysia and 5.1 Mha in Sabah and Sarawak (MPOB, 2012). Eighty-seven percent of which, are on mature area, and 13% on immature area (Yearbook of statistics, 2012). The oil palm tree is a perennial plant whose economic life last about 25 years. Tenera hybrid is an improved variety currently widely planted. There are about two fructifications per year. Optimal soil is well drained, deep (rooting depth is 40 to 50 cm) and fertile. The optimum growth temperature is 26° C and a relative humidity above 75%. Oil palm requires sun exposure of 165 hours per month. Oil palm is an intensive industrial crop and a monoculture. There is only one planting system practiced in Malaysia with a planting density of about 150 trees per hectares. The annual water requirement is 1300 mm (about 350L per tree). The harvest season is between July and September and usually begins 30 months after field planting. The peak of production of palm oil is between the 7th and 18th year of growth. Yields starts decreasing after the 18th year. Fruits take about five to six months to develop before they are ready for harvesting. The fruits are developed in large condensed infructescence and are called fresh fruit bunches (FFB). Economic life of a tree is approximately 25-30 years and after that, the trees are felled for replanting.

Primary residues, transformation processes, and secondary residues

Primary residues generated on the fields are trunks and fronds (above ground biomass). Significant quantities are available at each end of crop cycle (replanting). Some studies are available to estimate above ground biomass in oil palm fields. Lignocellulosic biomass residues that are produced include oil palm trunks (OPT), oil palm fronds (OPF), empty fruit bunches (EFB) palm pressed fibres (PPF) and palm shells.

Biomass palm oil mills use large quantities of water and energy and generate large quantities of solid waste, wastewater and air pollution. The solid wastes may consist of empty fruit bunches (EFB), mesocarp fruit fibres (MF) and palm kernel shells (PKS). The liquid waste is generated from the extraction of palm oil in a wet process in a decanter. This liquid waste combined with the wastes from cooling water and sterilizer is called palm oil mill effluent (POME).

The oil palm produces two types of oils, palm oil from the fibrous mesocarp and lauric oil (or kernel oil) from the palm kernel. In the conventional milling process, the fresh fruit bunches are first sterilized. The FFB are steamed in pressurised vessels up to 3 bars to arrest the formation of free fatty acids and prepare the fruits for subsequent processes. The sterilised bunches are then stripped of the fruitlets in a rotating drum thresher. The fruitlets are then conveyed to the press digesters. In the digesters, the fruits are heated using live steam and continuously stirred to loosen the oil-bearing mesocarp from the nuts as well as to break open the oil cells present in the mesocarp. The digested mash is then pressed, extracting the oil by means of screw presses. The press cake is then conveyed to the kernel plant where the kernels are recovered. In this step EFB waste can be collected. The oil from the press is diluted and pumped to vertical clarifier tanks. The clarified oil is then fed to purifiers to remove dirt and moisture before being dried further in the vacuum drier. Palm oil mill effluent (POME) is not exploitable for bio-jetfuel.

Based on the official data available, we estimated that Peninsular Malaysia

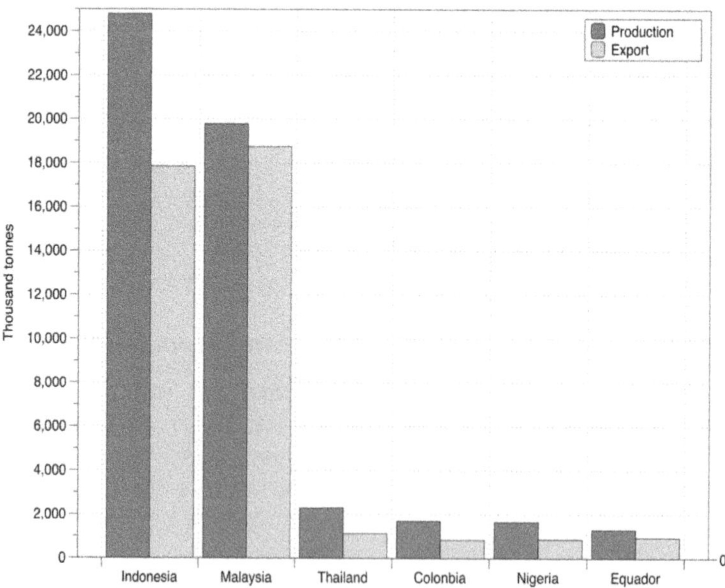

Figure 13: Major palm oil producers and exporters - source: (MPOB, 2012).

produces up to 17 million tonnes of oil palm trunks every year, along with respectively 7, 9, 5, and 2 million tonnes of oil palm fronds, empty fruit bunches, fibre, and shells.

Residues	Oil palm trunk			Oil palm fronds		EFB	Fiber	Shell
	Core	Middle	Outer	Leaflet	Rachis			
Moisture content (%)	83	75	68	57	70	60-50	40	20-10
RPR		-			-	0.230	0.140	0.065
Our residue estimation (Mt/year, fresh matter)								
Malaysia		35,25		14,10		19.2	11.7	5.4
Peninsula		17,80		7,12		9.4	5.7	2.6
Comparison with existing data from recent literature (Mt/year, fresh matter), data for Malaysia								
Sumathi S, 2008		8.2		12.9		15.8	9.6	4.7
MOPB, 2009		-		-		18	11	4.5
Bahari, 2010*		13.9		2.65		-	-	-
MiGHT 2013		6.63		46.53		22.43	-	5.61

Table 4: Estimation of oil palm residues and uncertainty of the data (Mt/year, fresh matter).

Nota bene: This estimation is highly imprecise, because the data is contradictory according to different sources, with methodologies of calculation of the sources, which may be disputable.

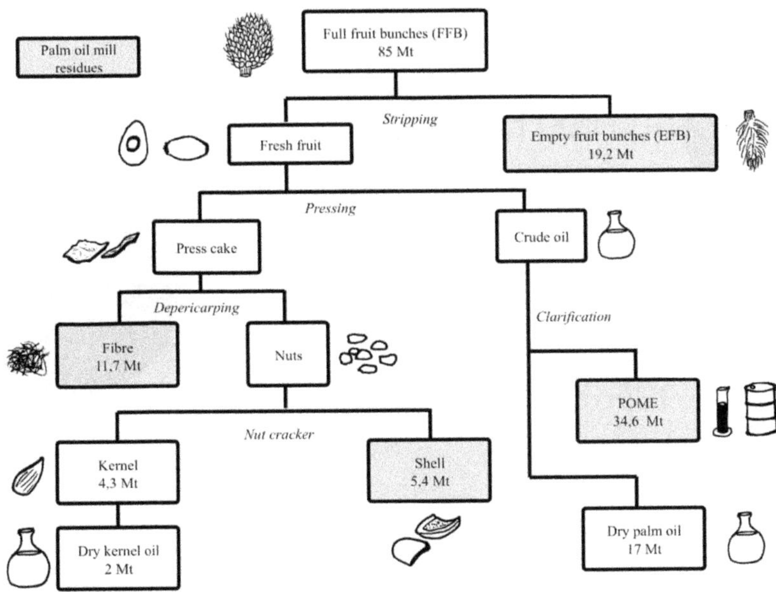

Figure 14: Oil palm mill processing flow chart in Malaysia.

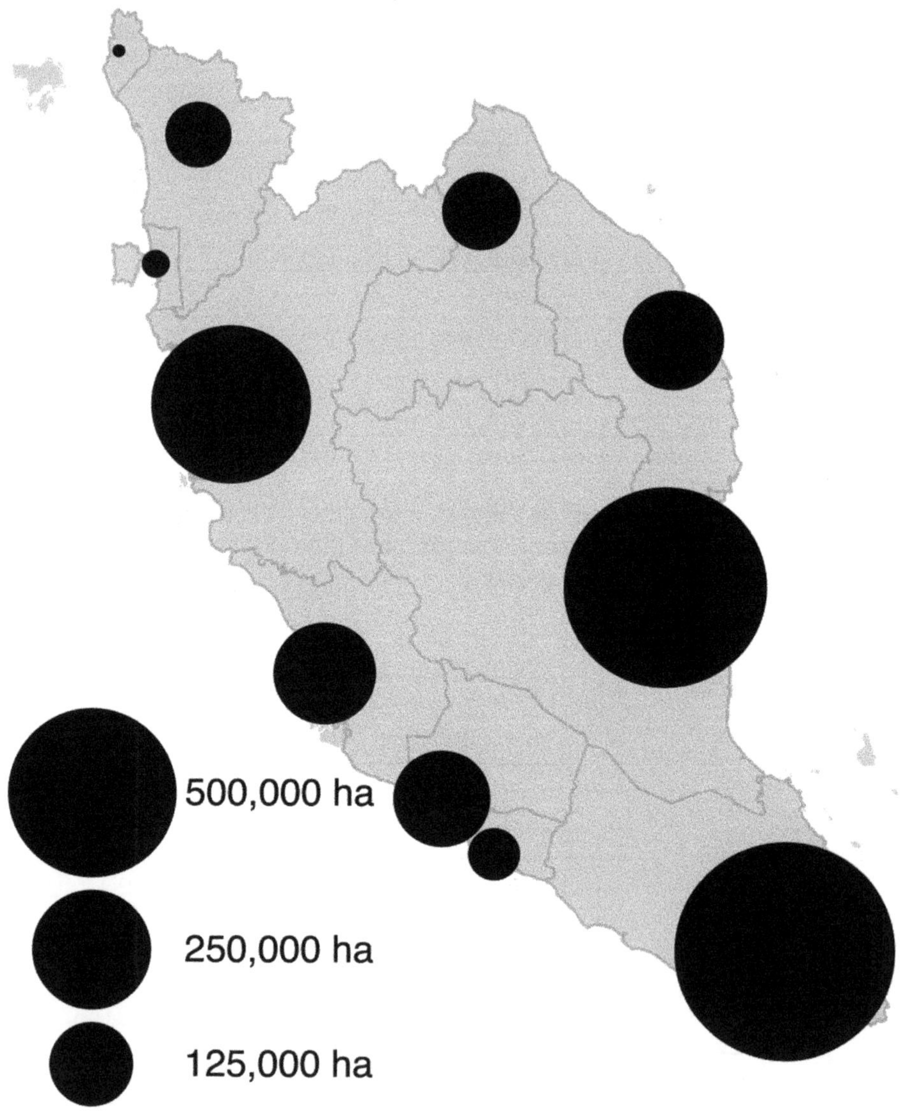

Figure 15: Distribution of palm oil plantations in Peninsular Malaysia (ha).

Chemical composition of residues

The lignin, holocellulose and alphacellulose content were reported to be 18.1%, 76.3% and 45.9% for oil palm trunk, and 18.3%, 80.5% and 46.6% for palm fronds respectively. The highest amount of lignin is found in the bark with 21.85% (UNEP 2012). Basically, the oil palm biomass contains is about 18-21% of lignin, and 65-80% of holocellulose (alphacellulose and hemicellulose), which is more or less comparable with other wood or lignocellulosic materials (Hashim et al. 2011 & Kosugi et al. 2010). The highest starch and total sugar contents are found in the core of the trunk. Total sugars were composed of glucose, xylose, arabinose and fructose with high values found in the trunk core with 6.55 mg/ml of glucose, 6.2 mg/ml of xylose, 1.31 mg/ml of arabinose and 0.04 mg/ml of fructose. From these values it can be concluded that the trunk could be an interesting resource material for sugars and starch.

% of dry weight	Trunk	Fronds	Bark
Lignin	18.1	18.3	21.85
Hemi-cellulose	25.3	33.9	58.95
Alpha cellulose	45.9	46.6	18.87
Ash	1.1	2.5	-

Table 5: Chemical composition of oil palm biomass (% of dry weight).

Oil palm component	Starch (mg/ml)	Sugars (mg/ml)				Total sugar (mg/ml)
		Glucose	Xylose	Arabinose	Fructose	
Bark	4.14	3.53	6.55	1.15	0.22	11.42
Leaves	2.53	2.17	3.79	1.7	-	7.66
Fronds	3.1	5.31	6.5	1.33	-	13.14
Mid-part of the trunk	12.19	5.97	6.61	1.09	-	13.67
Core-part of the trunk	17.17	6.55	6.2	1.31	0.04	14.06

Table 6: Starch and sugars content of different part of the oil palm.

Current use of palm residues

Oil palm tree can be used as lumber, pulp and paper producing materials, reconstituted boards and bio-composites. It has also been used as cellulosic raw material for panel production such as particleboard, medium density fiberboard or even block board. Even though oil palm wood can be used for panel production, oil palm-based plywood mills utilise only about 40% of the trunk leaving 60% to waste due to lack of desired properties (UNEP, 2012). The outer part is used for plywood, while the inner part is discarded. Furthermore OPT can be used for veneer lumber. This means that alternative resource from oil palm crops biomass waste, could fulfil the demand for alternative energy resource, and still provide substantial volumes for a sustainable second-generation biofuel.

So far a considerable quantity of tree trunks has been burned in the field, as there is often little demand for the wood in such rural areas. Fronds are normally left in the plantation as a mulching agent. However palm fronds and stems are currently underutilised. Moreover, these oil palm wastes have created a major disposal problem. Therefore, maximising energy recovery from the wastes is desirable for both environmental and economic reasons. When properly used, it will solve disposal problems and create value-added products. Therefore, the oil palm industry must be prepared to take advantage of this situation and utilize the available oil palm biomass in the best possible way, to convert waste to wealth. Palm oil mills in Malaysia are self-sufficient in energy. Fibre and EFB are burned as fuel in the boiler to generate steam.

4
Paddy biomass

Paddy

In 2012, 0.68 Mha of paddy were planted in Malaysia, among whom 0.50 Mha in the peninsula. Most of the paddy fields are wetland paddy. Only 0.07 Mha is dryland paddy. Paddy lands are mainly located in Kedah, Perak, Kelantan and Perlis states (Yearbook of Statistics, 2012). Between 2008 and 2012, the average production of paddy was 2.5 Mt in Malaysia. (Calculated with data from MDA). For Peninsular Malaysia the average production of paddy on this

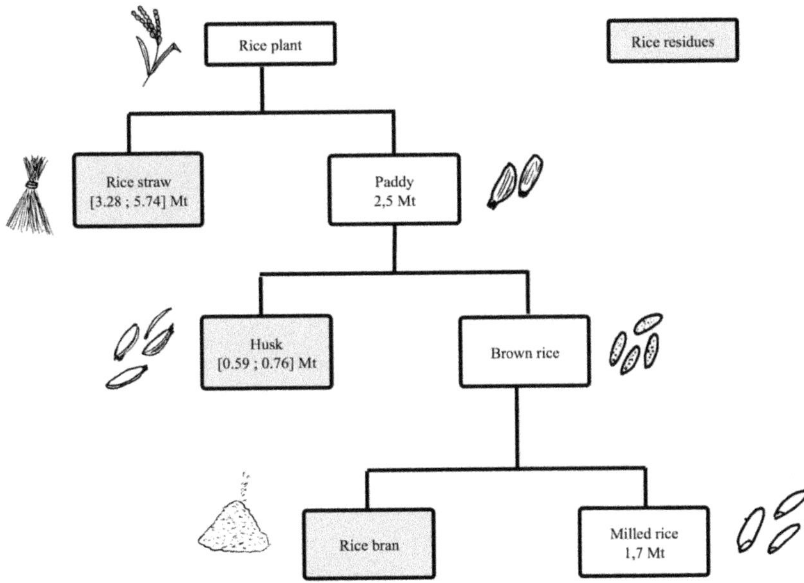

Figure 16: Paddy - rice flow chart in Malaysia.

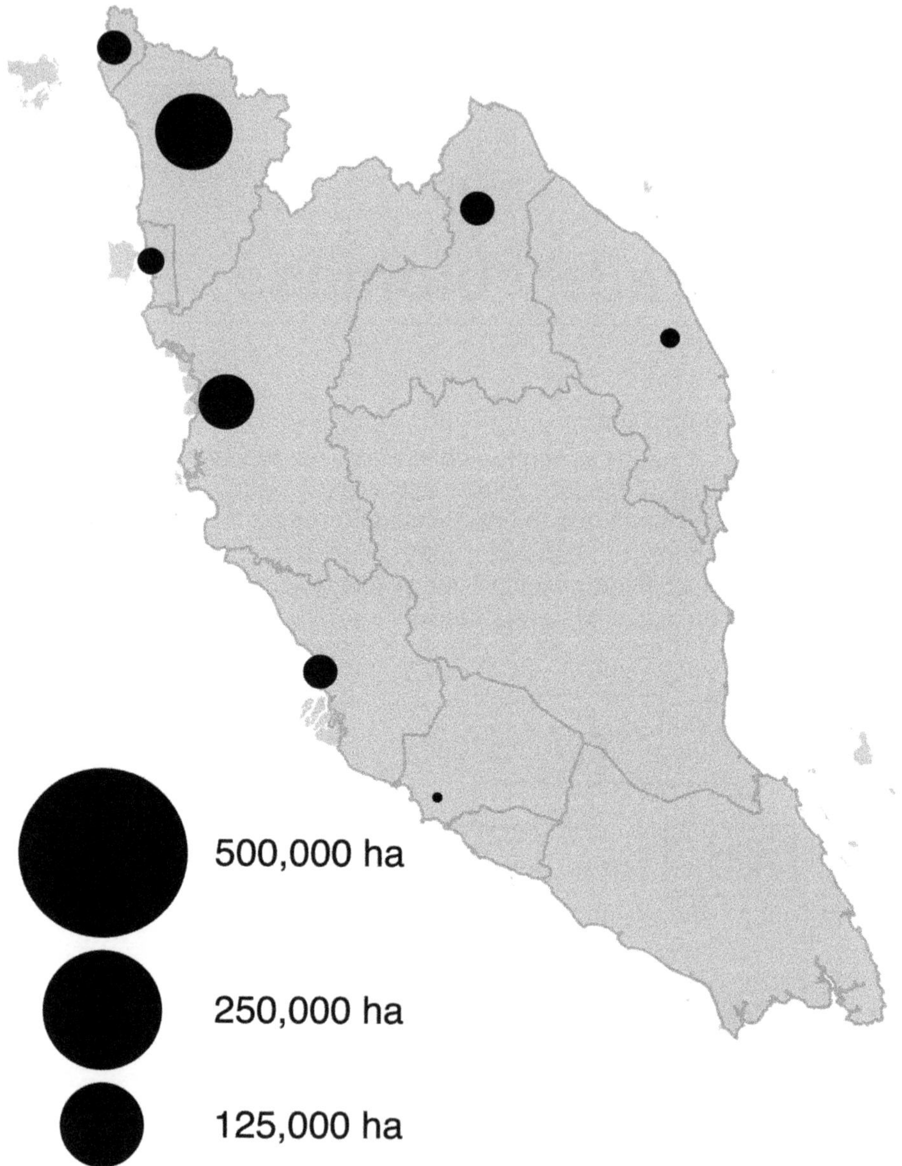

Figure 17: Distribution of paddy fields in in Peninsular Malaysia (ha).

average production of paddy was 2.5 Mt in Malaysia. (Calculated with data from MDA). For Peninsular Malaysia the average production of paddy on this period was 2.16 Mt. In Malaysia the number of harvest seasons varies from one to two per year. A rice plant can produce until 30 stalks; roots are between 40cm to 1m deep. The average yield for wetland paddy in Peninsular Malaysia between 2008 and 2012 was 4,199 kg/ha. Yield is lower for drylands paddy (865 kg/ha in Malaysia - 2008). There is a majority of smallholders among 296,000 paddy farmers with 65% of them having less than one hectare. The average size farm is 1.06 ha. In Malaysia, mechanised paddy farming dominates but many traditional systems are still practiced.

Primary residues, transformation processes, and secondary residues

Paddy straw is the main primary residue of generated by paddy cultivation. Rice straw Residue-to-Product-Ratios (RPR) values range from 0.4 to 3.9, based on the different practice of harvesting with an average moisture content of 12.71%. Given the variability of cultural systems in Malaysia, the uncertainty and variability of RPR values, a certain degree of uncertainty on the calculation for the potential quantity of residues is unavoidable. Rice husk is the major by-product of the rice milling industry and the most commonly available lignocellulosic material that can be converted to different types of fuels and chemical feedstocks through a variety of conversion processes. Husk represents 20 to 22 % of the paddy weight (Chouragade, 2012).

Chemical composition of residues

Rice straw is a by-product of rice production and is a great bio resource. Rice straw predominantly contains cellulose 32-47%, hemicelluloses 19-27%, lignin 5-24% and ash 18.8%. The carbohydrates of rice straw include glucose 41-43.4%, xylose 14.8-20.2%, arabinose 2.7-4.5%, mannose 1.8% and galactose 0.4% (Vasan, 2012).

Current use of paddy residues

Rice residues are an important source of energy for both domestic as well as industrial purposes. Most of the husk from the milling is either burnt or dumped as waste in open fields and a small amount is used as fuel in the rice mills to generate steam for the parboiling process, electricity, as a component of cattle feed or bulking agents for composting of animal manure. Very often, the bulk of the straws is left rotting in the field or is ploughed back into the soil as a conditioner and organic fertiliser.

Main analysis	Rice bran	Rice straw	Rice husk
Moisture content (% of fresh matter)	10	7.2	8.1
Crude protein	14.2	4.2	3.7
Crude fibre	4.1	35.1	42.6
NDF[1]	12.4	69.1	67.8
ADF	3.2	42.4	51.7
Lignin	1.2	4.8	14.2
Ether extract	13.2	1.4	1.5
Ash	6.9	18.1	5.3
Starch	42	-	17.5
Total sugars	3.8	-	-
Gross energy, (MJ/kg, dry matter)	20.5	15.5	16.3

Table 7: Chemical composition of paddy residues.

	Production (Mt/year, fresh matter)		
	Paddy	Rice straw	Rice husk
Malaysia	2.5	[3.28 ; 5.74]	[0.59 ; 0,76]
Peninsular Malaysia	2.16	[2.83 ; 4.96]	[0.51 ; 0,67]
Comparison with existing data for Malaysia from recent literature (Mt/year, fresh matter)			
MDA, 2009	-	0.88	-
Abdel-Mohdy et al., 2009	-	-	0.48
MiGHT, 2013	2.82	2.82	0.39
Moisture content (%)	14	11-13	8-9

Table 8: Estimation of paddy residues and uncertainty of the data (Mt/year, fresh matter).

5
Rubberwood biomass

Rubber

Rubber trees currently cover around 0.770 million ha in Malaysia. This situation is the result of a progressive decrease of the areas during the last decade. The total area was above 1,200 million hectares in 1998 and decreased to 0.99 million ha in 2012. Rubber trees are a perennial tree crop. Its economic life period is about 25 years. Rubber is adapted to equatorial lowlands with well-drained soil. In marshy area with poor physical properties and waterlogged conditions, growth of rubber is very weak. In Malaysia, rubber trees are supposed to be harvested after 25 years on average, when they are too old to produce latex at an economic rate. The mature trees are then felled. At replanting, large trees of economic value should be removed first followed by felling and removing of smaller trees and slashing of the under-growth vegetation. The replantation rates depends on the equation between rubber profitability compared to other crops, such as oil palm, and finally become “rubber activity” following the equation: replantation plus new plantations minus what is replanted with a different crop. With an average of 25 year lag between the plantation and the harvest, the supply of rubberwood logs every year depends on how much hectares were planted 25 years ago, plus or minus a few years of adaptation by the smallholders (indeed, according to the fluctuation of latex prices, the farmers can decide to harvest their trees sooner when the latex price is too low, or to harvest later when the latex price is very good, making it more profitable to wait a few years more). Nowadays the sector is under intense pressure as it is at its lowest level ever of mature trees, but a wave of plantations done in the recent years is expected to change the situation in the coming years, with the maturation of all the young plantations (Roda et al., 2012).

Primary residues, transformation processes, and secondary residues

The only relevant lignocellulosic residue created in the rubber sector is the rubberwood. It originally is a residue of the latex business, but for the past two to three decades, rubberwood has become an important source of raw material for the panel and furniture industry. When the old trees are felled, before replanting, the trunks and the biggest fraction of logging residues are incorporated in the transformation sector. As Ratnasingam, showed (2012), the MDF sector already has integrated the use of milling residues in their manufacture (see Figure 18). Today the supply of rubberwood responds to cyclical dynamics, which generates tension in the market. Between 2010 and 2011, it was estimated that the potential supply of rubberwood was 4 Mm3 in Peninsular Malaysia (Roda et al., 2012). The supply is expected to increase until 2030 – 2035 before another decline. Further prediction on rubberwood availability will depend on current trends on replantation (i.e., policies, subsidies, fluctuation of latex prices, demand of rubberwood furniture, etc.). Industrial projection for bio-jetfuel based on rubberwood residues in such an uncertain context seems limited. Thirty-five percent of the above ground biomass from rubber smallholdings is often left behind due to logistics and transportation reasons. (Ratnasingam and Jones, 2011). Further downstream

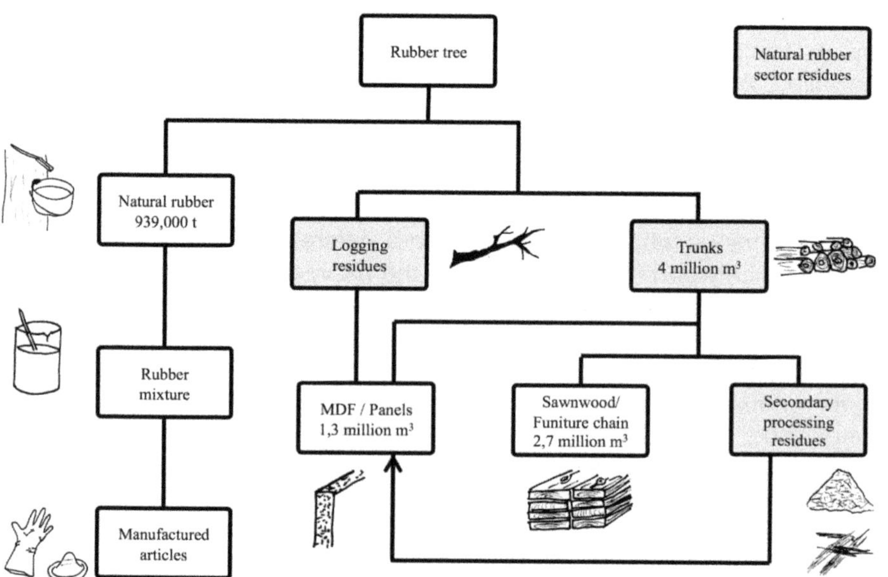

Figure 18: Flow chart for rubber sector in Malaysia, and current estimation from Roda, 2011 (Malaysian Timber Council unpublished report).

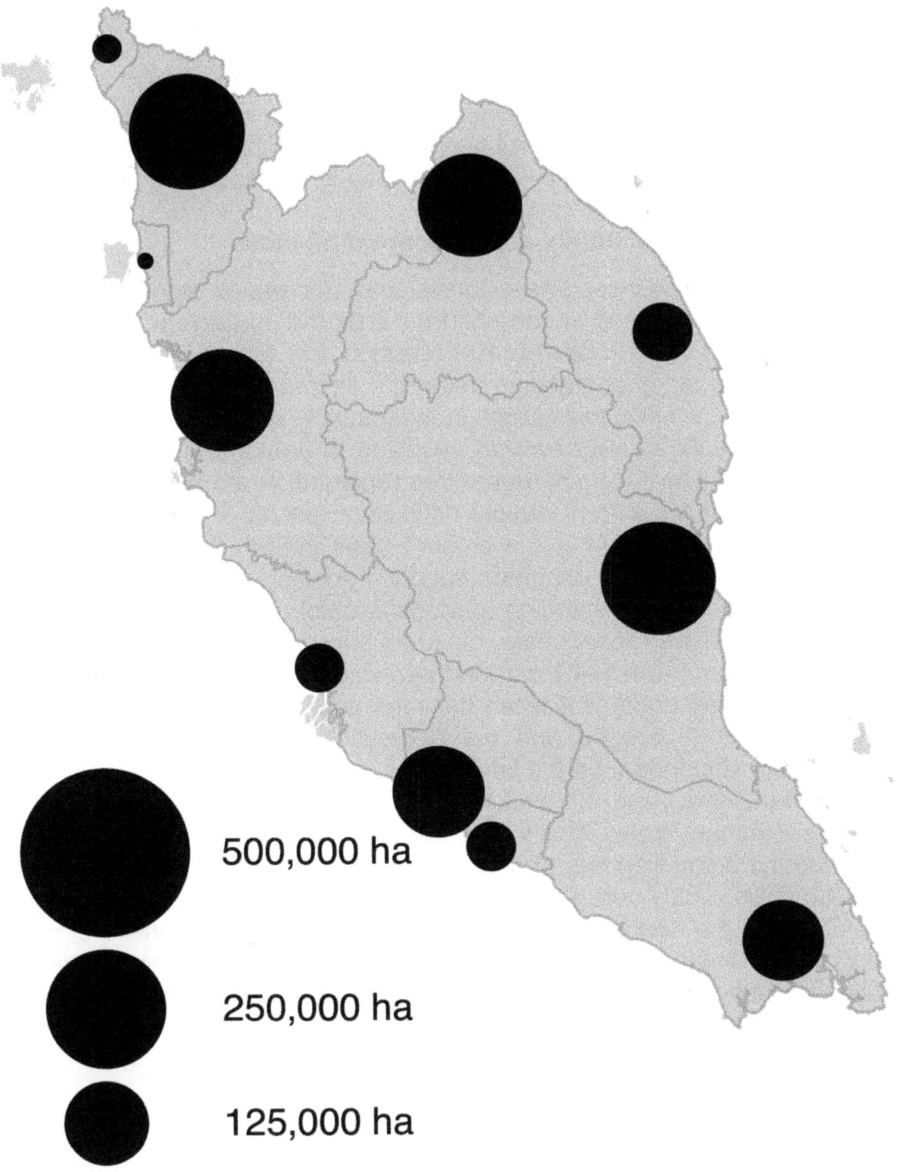

Figure 19: Distribution of rubber plantations in Peninsular Malaysia (ha).

in the rubberwood sawmills and furniture industry, it is estimated that rubberwood could generate 1.484 million m3 of wood residues in Malaysia, of which 1.162 million m3 is in the peninsula, while Roda et al., 2012 estimated that only 0.3 million m3 of potential lignocellulosic waste is not already used. The chemical composition of rubberwood falls within the variability in composition of wood species, see the forest resource chapter.

Fluctuations of the availability of rubberwood biomass

The total area of rubberwood has consistently decreased since the 1980s. Replantation has decreased in general (from 0.07 million ha/year in the 1960s to 0.02 million ha/year in 2010). Rubberwood log prices have increased proportionally. The stock of standing trees has decreased in some states and increased in some others: this results in local supply disparities. Replantation programmes have been very cyclical, resulting in peaks and lows in supply (25±4 years later). Low levels of rubber tree replantation around 25 years ago, is the main reason of current supply difficulties: peaks of past replantation determine today's supply. After the present crisis period, the next few years will temporarily witness a much better supply. Peninsular Malaysia has a total of around 4 million m3 of standing sawlogs & chiplogs within a year across 2010-2011. Only in the recent years, did latex price become a strong factor in the supply situation. From 1949 to 2000 there was no real connection between hectarage and latex price. But when the total area fell below 1.3 million ha, latex price (SMR20) soared and the domestic price of rubberwood logs became directly dependant on the latex price.

Over the years, the trend of uptake has consistently been dictated by the availability of mature trees. The logs consumption by MDF & particleboards reaches around 3.1 million m3, while logs consumption in the furniture chain is around 0.6 million m3 / year (estimation between September 2010 to August 2011). Despite the tension on the rubberwood market, there are more rubberwood volumes theoretically available (for sawn timber) than what the furniture industry is actually consuming. The gap is explained by lack of consolidation of the supply along the value chain, which makes the supply cost higher than expected, despite the good road network.

	Sabah and Sarawak	Peninsula
Rubberwood offer (Million m3) [1]		4
Rubberwood residues (Million m3) [2]	0.3	0.3-1.1
Rubberwood residues (Mt, at 12% humidity; density = 650 kg/m3)	0.2	0.2-0.7

Table 9: Potential and actual availability of rubberwood according to Roda, 2012 [1] and Ratnasingam, 2012 [2].

6
Sugarcane biomass

Sugarcane

There are only 45,000 ha in Peninsular Malaysia (FAOSTAT, 2011), located in a few states in northern peninsula due to climatic requirements. A dry season is essential for good plant growth. The crop is absent in Sabah and Sarawak. Sugar cane gives a very high dry matter yield per unit of land area (Koopmans and Koppejan. 1997) and possesses a high valorisation potential for bioenergy in the countries where it is planted in large quantities. Between 2008 and

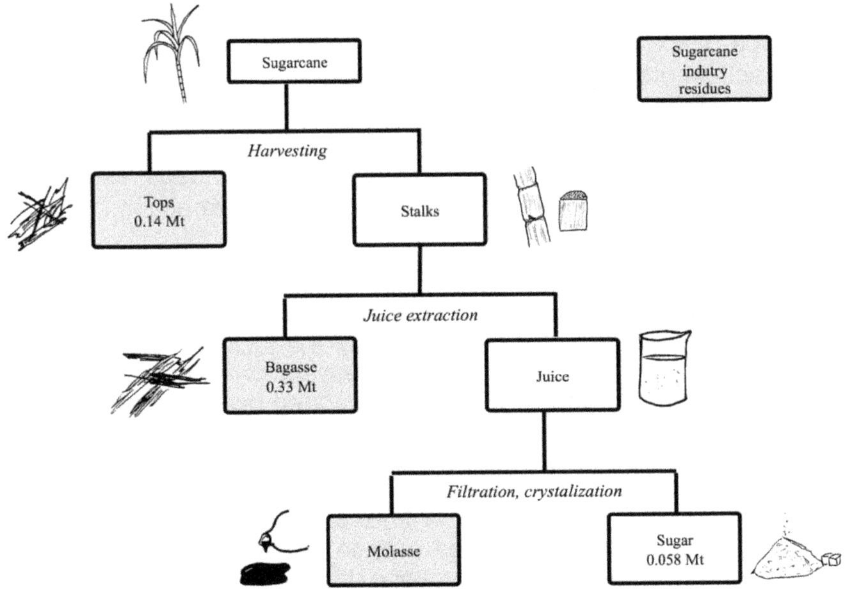

Figure 20: Sugarcane industry flow chart in Malaysia.

2013, the average annual production in Malaysia was 0.825 Mt/year. (FAOSTAT). The lack of sugarcane development in Malaysia reflects the higher remuneration received by farmers from other crops, especially oil palm. Large estates represent 85% of the sugarcane area with 60t/ha per year of cane stalks (that is 75% of the entire fresh weight). Smallholders represent 15% of the area, yielding 40t/ha. The water requirements reach 150mm/month. The plant need water during the growth period (first 8-9 months) and during the dry season of 4 months, watering is required at least every 15 days. Sugar cane is a perennial plant that requires tropical climate for optimal growth. Harvests take place during the dry period, between January and April in the west coast of Malaysia. Optimal period is when the sugar is accumulated in the stalks until it reached the maximum at maturity. Harvest can take place nine to twelve months after planting or after regrowth. The first step is to cut the stalks leaving the lower part (the strain) to allow the plant to regenerate. Manual harvesting can obtain 3-5 tonnes per day while mechanisation increases the rate to 60 tonnes of stalks per day. Fields are replanted every 5 to 10 years (CIRAD, 2014).

Primary and secondary residues

Traditionally large quantity of residues remain on the field after harvest. The major components of sugarcane residues on the field are straw, green leaf, sheath, tops, stalk including the presence of mechanical impurities such as roots and soil. In order to facilitate subsequent works, it is possible to burn the field but this practice is decreasing. The material tends now to be collected from the ground after mechanical harvesting. Conversely, bagasse comes from industrial processes involving repeated extraction steps. Bagasse is the fibrous by-product of sugarcane stalks milling for juice extraction.

Current use of sugarcane residues

Malaysia has four sugarcane processing facilities, one each in the states of Perlis, Kedah, Penang, and Selangor. Two of the facilities, Gula Padang and Perlis Plantations are integrated mills processing cane into raw and refined sugar with the added capability of refining imported raw cane sugar. The other two facilities are refineries handling imported raw cane sugar. One is across from Penang Island, and the other is located near Kuala Lumpur in the state of Selangor. Since Malaysia's sugar processing industry depends on imports for about 90% of its raw sugar, with the bulk being processed at Penang and Kuala Lumpur, bagasse is produced only in the two facilities in Perlis and Kedah. Bagasse, as well as sugar cane tops and leaves and sugarcane straws may be used as an energy source for steam generation, mulching

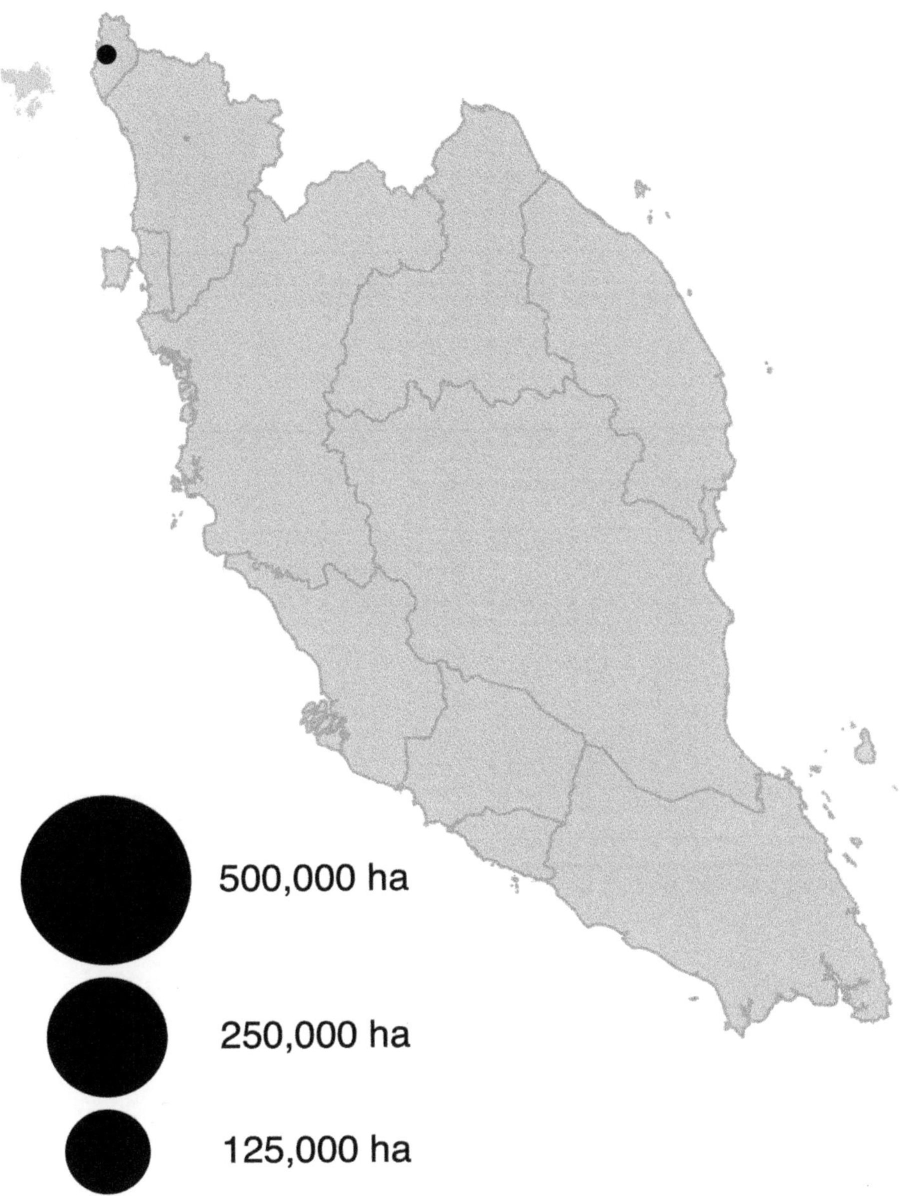

Figure 21: Distribution of sugarcane plantations in Peninsular Malaysia (ha).

agent, cattle feed or are burnt onto the field. Most sugar factories burn all the bagasse they generate, even at very low efficiencies. This is done to ensure that all bagasse is burned, as dry bagasse is known to be a fire hazard. In some countries bagasse is also used as a raw material for the paper and board industry. Worldwide as in Malaysia, the interest for second-generation biofuels based on sugarcane lignocellulosic residues is now becoming a real option.

Main analysis	Sugarcane trash[1]	Bagasse
Dry matter (% of fresh matter)	40 – 45 % (Tops) / 85 % (Dried leaves)	50 % [3]
Cellulose	40.1 % (± 0.4)	41.8 % [2]
Hemicellulose/Polyoses	30.7 % (± 0.2)	24.9 % [2]
Lignin	22.9 % (± 0.2)	23.2 % [2]
Ash	2.2 % (± 0.2)	1.9 % [2]
Extractives	3 % (± 0.3)	6.8 % [2]

[1] (From Gomez et al., 2014, % of weight dry basis) [2] (Average from literature review in (Canilha, 2012), % of weight, dry basis) [3] (Anwar, 2010)

Table 10: Chemical composition of sugarcane residues.

Reference	Sugar Cane tops/leaves		Bagasse	
	RPR	Moisture content (%)	RPR	Moisture content (%)
Vimal '79	0.1	75	0,33	48
Webb '79	-	-	0.289	52
AIT - EEC '83	0.125	50	0.141	50
Strehler '87	-	-	1,16	40-60
Bhattacharya ea' 93	-	-	0.29	49
Ryan ea '91	-	-	0,2	-
USAID '89	0.3	-	-	-
Average	0.175	62.5	0.402	49.3
Estimated production of residue (Mt/year, fresh matter)				
Peninsula*	0.14		0.33	

Table 11: Sugarcane residues (* All sugarcane crops of Malaysia are located in Peninsula).

7
Coconut biomass

Coconut

The total area of coconut plantation is about 0.11 Mha in Malaysia among which 0.065 Mha are on the Peninsula (MYS, 2012). The major states growing coconut are Perak (15,180 ha), Johore (21,250 ha), Selangor (10,320 ha), Sarawak (22,290 ha) and Sabah (19,150 ha). Between 2001 and 2013, the average production of fresh coconut was about 0.52 Mt per year. (average from MARDI - Department of Agriculture Malaysia). Coconut palm (Cocos nucifera) is a perennial plant able to grow in a wide range of soil types. Coconut palm begins to bear fruit after 7-10 years of cultivation and remain productive for over 100 years. Coconut in Malaysia is a smallholder's crop with 91% under smallholder cultivation and 9% under estate management with an average coconut farm size of 2.8 ha and a production yield varying from 3,000 to 8,000 nuts per hectare producing about 93% of the total coconuts in the country. Of the total area planted, 63% is located in Peninsular Malaysia, 19% in Sabah and 18% in Sarawak (FAO 2013). With increasing labour shortage, decreasing productivity of palms and the massive conversion of coconut lands to oil palm plantation and other more profitable crops, the country projects a continuous decline in coconut plantation area. However, coconut still plays an important role in the country's economy providing livelihood to about 100,000 farmer families or in other terms to about 10% of the nation's farming community (Yahya et Zainal, 2014). Most varieties mature after 15 years and then offer a steady nut production for decades. The fruit matures in 12-14 months after fertilisation of the flower. When the nuts are ripe they come off the bunch and fall to the ground where they are collected. Direct collection of coconuts depends on whether the consumer demand is for young coconuts or for matured old coconuts (coprah).

Primary residues, transformation processes and secondary residues

There are three categories of primary residues from coconut cultivation, i.e., fronds and debris that are shed throughout the year, and wastes generated only at the end of the plantation life, when trees are felled before replantation. Coconut plantations are quite scarce and highly scattered in Malaysia, and the rotation is much longer than in oil palm plantations. Therefore primary residual biomass could hardly constitute an effective opportunity for bio-jetfuel.

Current use of coconut residues

Coconut shell is used to produce charcoal. It is extensively used for the manufacture of activated carbon due to its high absorption capacity. Shell charcoal is then transformed into activated carbon, which has the ability to absorb effectively trace quantities of either unwanted or valuable liquids or gases. Activated carbon is used in solvent recovery processes, water and effluent treatment. Moreover 30% of the coconut husk is fibre and the leftover 70% is coir dust (ICAR, 2013). Coir pith is used as manure (after composting), mulch material and to make briquettes. The coir pith briquettes can be used as a substitute fuel in the place of firewood for the tile and brick industries. The coconut wood is used for making wall panels, and furniture's. Coconut leaves

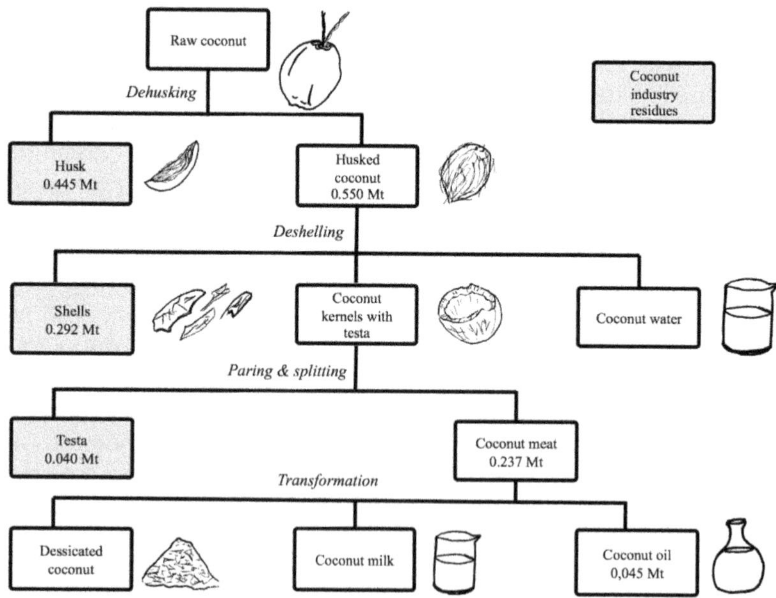

Figure 22: Coconut industry flow chart in Malaysia.

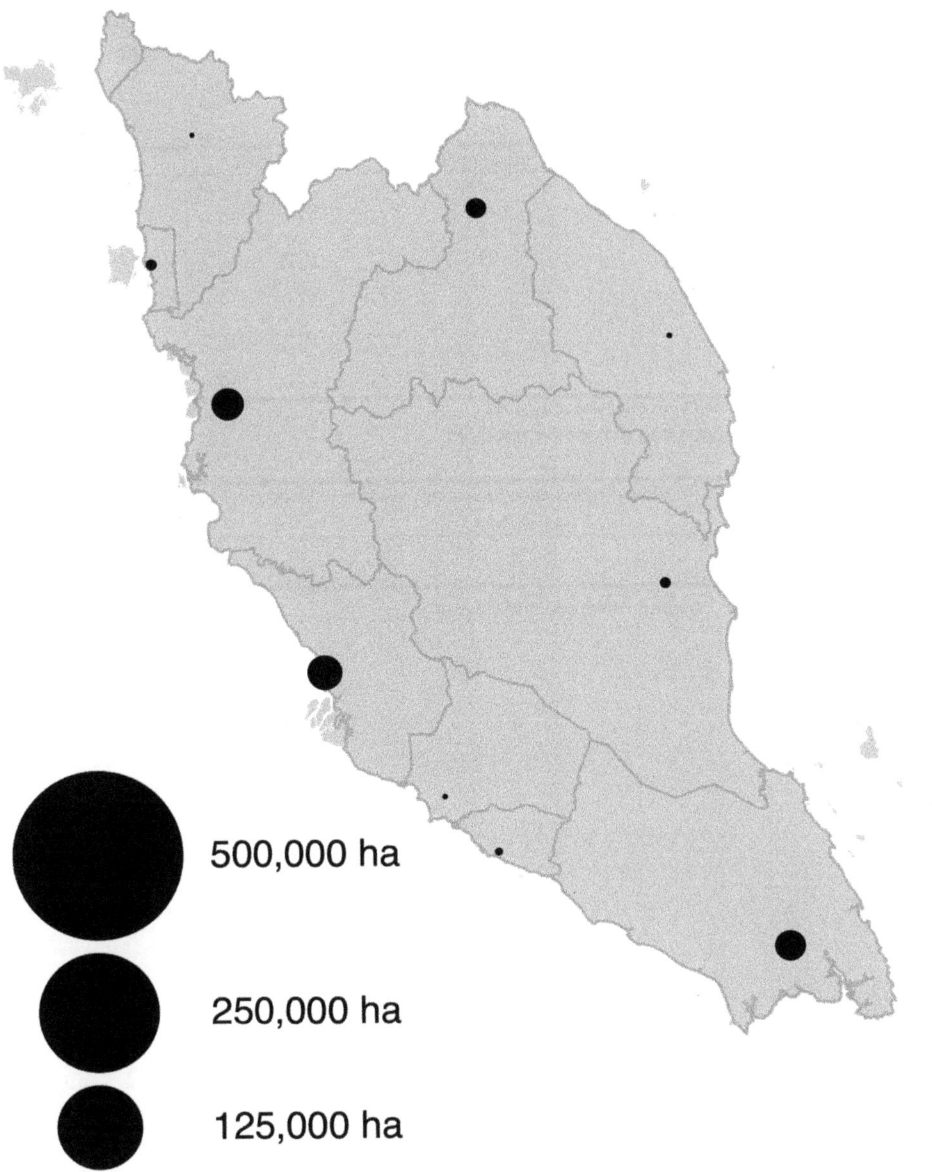

Figure 23: Distribution of coconut plantations in Peninsular Malaysia (ha).

are plaited for thatching houses and sheds. They are also used for making baskets, temporary fences, etc. Midribs of the leaves are used to make brooms.

Main analysis	Coconut husk	Coconut husk	Coconut
Moisture content (% of fresh matter)	15	20 – 15	11.1
Cellulose	23.87	24.25	-
Hemicellulose	8.5	-	-
Lignin	29.33	44	-
Ash	-	6.19	2.28
Crude fibre	-	-	32.39
Crude fat/oil	-	-	2.14
Protein	-	-	0.46
Carbohydrate	-	-	52.63

Adapted from Tejano, 1985, and from Ewansiha et al., 2012

Table 12: Chemical composition of coconut residues.

Source	Estimated production of coconut residues in Malaysia (Mt/year, fresh matter)	
	Coconut Husk	Coconut shell
This study	0.445	0.292
Bilba et al. 2007	0.171	-

Table 13: Estimation for coconut residues in Malaysia.

8
Woody biomass

Forestry sector

There are 12 million ha in Sabah and Sarawak, and another 5.8 million ha of forest in Peninsular Malaysia. In the Peninsula, only 3 million ha are devoted to production, and this area is entirely ecocertified under PEFC. There are four types of forest residue that can be derived from the forest sector:

- Logging residues: bark, stumps, tops, branches.
- Sawmilling residues: sawdust, off-cuts, slabs, shavings and bark.
- Plywood and veneer residues: veneer cores, defective ends, and irregular pieces of veneer sheets.
- Secondary processing residues: sawdust, plane shavings, small pieces of timber trimming, edging, bark and fragments.

Most of the wood residues are left in the forest to rot, especially in sparsely populated areas where demand for wood fuel is low. However, forest residues also play an important role in soil fertility and the total removal of all above ground residues can accelerate soil degradation. Logging residues consist of 12% stem-wood above first branch, 13.4% branch wood, 9.4% natural defects, 1.8% stem-wood below first branching, 1.3% felling damage, 1.6% stump wood and 0.5% other losses (FAO Forestry Department, 2013).

Wood transformation and secondary residues

Mill residues consist of two main groups. The first is made up of large pieces, bulk waste, while the second group consists of fine wood particles such as shavings, saw dust and sander dust. Recovery rates vary with local practices as well as species. Sawmill recovery rates range from 42 to 60% with an average of 51.6%. An ITTO study reported a recovery rate of 52% in the state of Terengganu after collecting data from 24 mills representing about 70% of the production in the state. Recovery rates are dependent on log diameter and size, and type of machines used.

In the round form, the logs contain about 12% waste in the form of bark. Slabs, edgings and trimmings amount to about 34% while sawdust constitutes another 12% of the log input. After kiln-drying the wood, further processing may take place resulting in another 8% waste (of log input) in the form of sawdust and trim end (2%) and planer shavings (6%) (Ratnasingam and McMullen, 2014).

Wood residue potential amounts to 13.98 Mt of fresh biomass in Malaysia, but with a uncertainty ranging from 6 to 16 Mt for Malaysia.

Hardwoods, which constitute the major share of Malaysian woods, contain 40-50% cellulose, 20-25% lignin, and 25-35% hemicelluloses.

Plywood and veneer production residues

Recovery rates vary from 45 to 50% with the main variable being the diameter and quality of the log. Of the log input, the main forms of waste are: log ends and trims (7%), bark (5%), log cores (10%), green veneer waste (12%), dry veneer waste (8%), trimmings (4%) and rejected plywood (1%). These form the largest amount of waste while sanding the plywood sheets results in another loss of 5% in the form of sander dust.

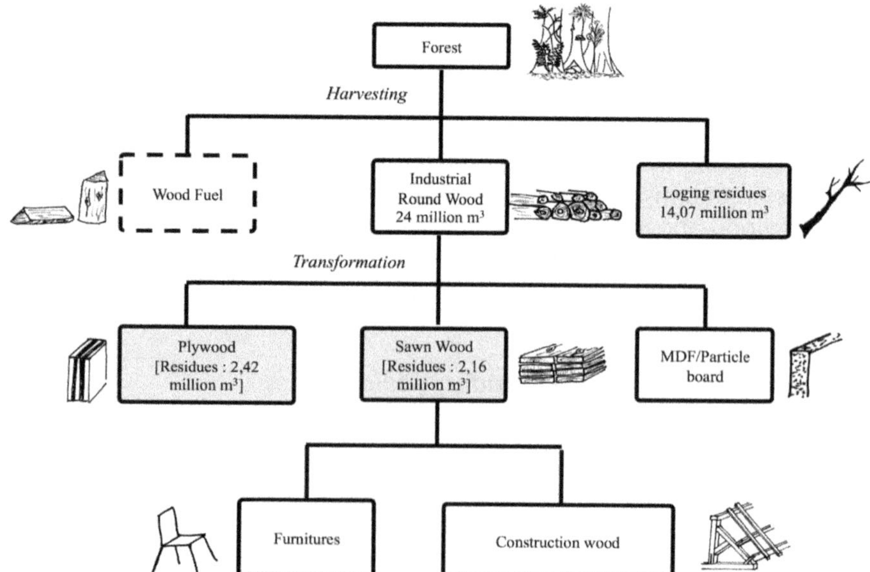

Figure 24: Flow chart for forest and wood product sector in Malaysia.

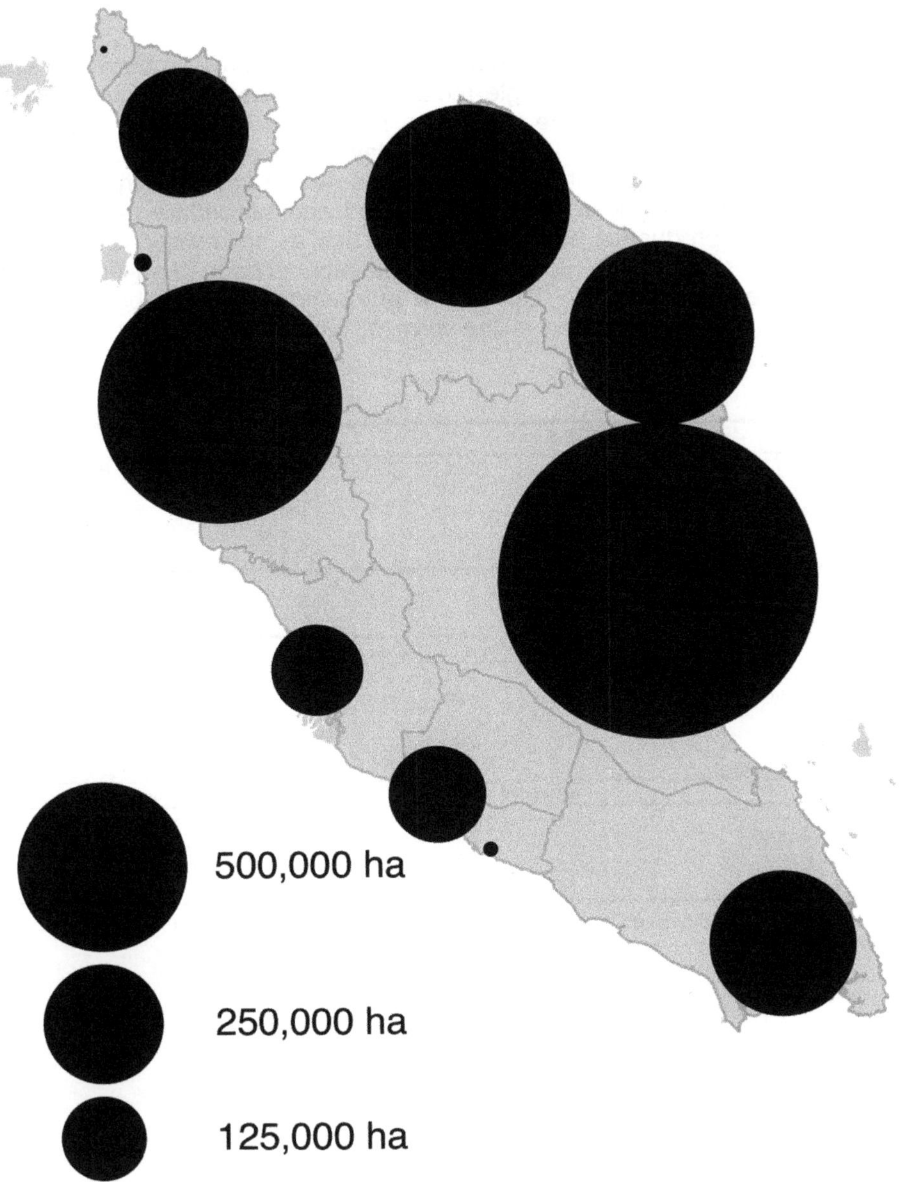

Figure 25: Distribution of forest in Peninsular Malaysia (ha).

Current use of wood residues

A significant share of mill residues is already used for energy production under cogeneration. In Malaysia the single most important use of wood residues is as fuel. In sawmills, most of the secondary manufacturing residues are used as fuel in wood-fired boilers. Peninsular Malaysia nevertheless continues to generate a significant amount of wood residues, which are not currently used for any downstream process. All types of wood and secondary raw material are used for the production of particle board such as solid wood, solid wood residues (off-cuts, trimmings), low grade waste such as hogged sawmill waste, sawdust, planer shavings, etc. During the production process about 17% to 40% of residues are generated in the form of trimmings, but this amount is recycled.

	Softwoods		Hardwoods	
% of dry weight	Wood	Bark	Wood	Bark
Lignin	25-35	40-55	18-25	40-50
Polysaccharides[1]	66-72	30-48	74-80	32-45
Extractives	2-9	2-25	2-25	5-10
Ash	0.2-0.6	Up to 20	0.2-0.6	Up to 20

Table 14: Chemical composition of wood.

Peninsular Malaysia	Estimated residues (Mt/year)	Uncertainty (Mt/year)
Logging residues	1.83	1.83-2.649
Sawntimber residues	0.97	0.97-1.16
Plywood residues	0.17	0,17-2.492
Veneer residues	0.02	-
Total Peninsular Malaysia	2.99	2.97-5.3
Malaysia	Estimated residues (Mt/year)	Uncertainty (Mt/year)
Logging residues	10.55	10.55-11.369
Sawntimber residues	1.62	1.62-1,81
Plywood residues	1.81	1.81-4.132
Total Malaysia	13.98	6-16

Table 15: Potential for forest and wood industry residues in Malaysia, in million tonnes / year (fresh matter).

9
Sustainable feedstocks for bio-jetfuel

All the six main sources of biomass in Malaysia produce substantial quantities of residues that are only partly used. This leaves room for sustainable feedstocks, provided that life cycle analyses confirm for each the perimeters of sustainable use versus unsustainable. Two major issues still remain. All the quantities are subject to considerable uncertainty, because of the data source. Firstly, official data often contradict themselves, differing from one source to another. Sometimes in the same report, different numbers are given at different pages, for the same biomass. Secondly, most of the commodities in question are already established in markets where the net value of the raw material makes them uneconomical for energy use except by valorisation of the residues or co-products. These residues and co-products are far from being completely used today. The potential for their effective valorisation will depend of several factors: competition with other possible valorisation, transport, preconditioning of the residues, location of the resource and the collection points and potential locations of bio-refineries.

Finally, the most important point is the fact that the largest pool of biomass seems to be residues of palm oil. These residues are already valorised for some part and recent technological advances as the recycling of oil palm trunk for composite wood and other higher value products may drastically change the real availability of these residues. But palm oil biomass is also constantly under attack by environmental groups and analysts, because of the link between oil palm plantation extension and deforestation. Taking palm oil biomass into consideration for bio-jetfuel generation can be a real question for some end-users, such as airlines, due to the sensitivities of their respective markets. Whether to take into consideration oil palm biomass or not, changes drastically the perspective. With oil palm biomass, the total potential biomass availability would sum up around 100 million tonnes of fresh matter in all Malaysia, or only 50 million tonnes in Peninsular Malaysia. Conversely, without

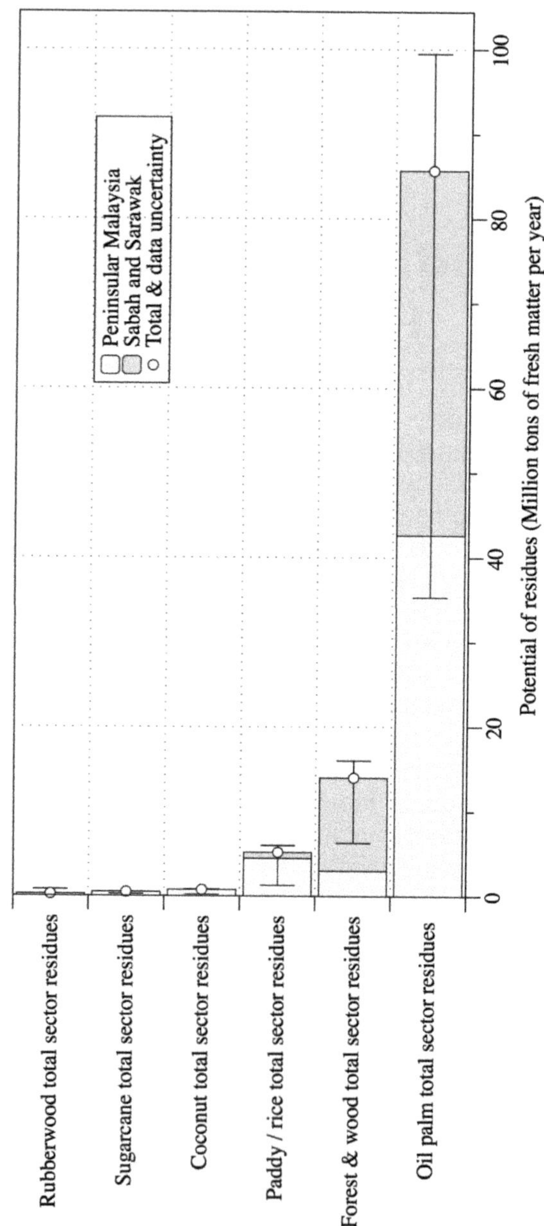

Figure 26: Potential for lignocellulosic resources from the major Malaysian crops.

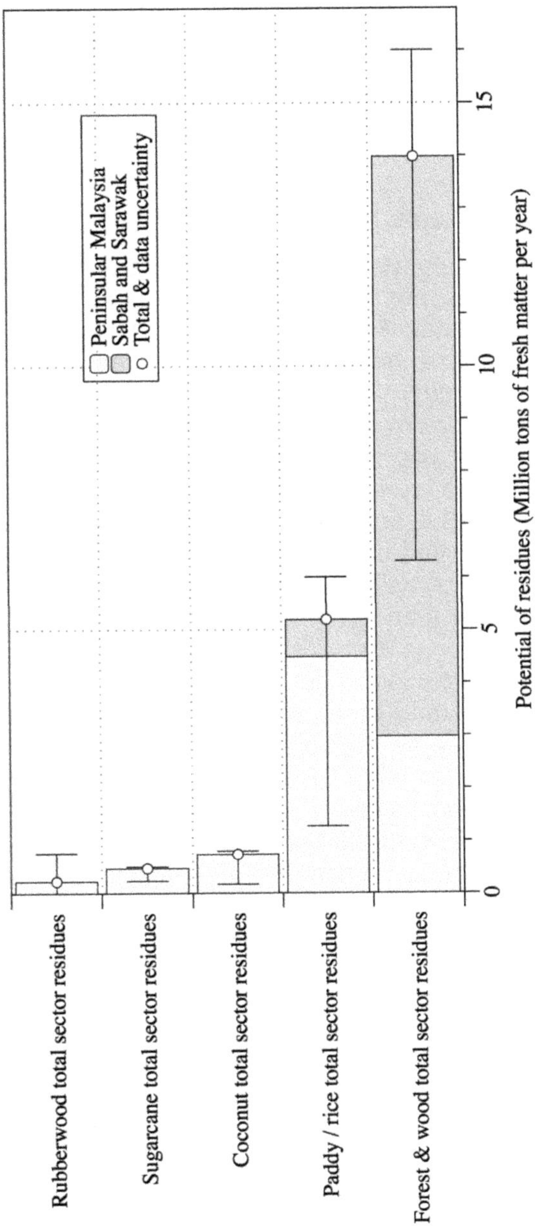

Figure 27: Potential for sustainable lignocellulosic resources from the major Malaysian crops, excluding oil palm resources

oil palm biomass, all Malaysia would offer 20 million tonnes of fresh matter, among which 7-8 from the peninsula. These indicative numbers are subject to uncertainties of more than plus or minus 50%. Note also that in Peninsular Malaysia, the potential of available paddy residues might be significant and of the same order as that of the available forest products residues, given the lack of precision of the data.

Theoretical maximum bio-jetfuel production

Published bio-jetfuel conversion efficiency vary around 8.6 kg of dry lignocellulosic biomass per kg of Jet A-1, according to existing aviation-certified processes (Matas Güell et al., 2012). Optimistic approaches project 6.1 kg of dry biomass per kg of Jet A-1 (unpublished Airbus Group communication, this study). Based on the latest, Malaysia would be able to produce a theoretical maximum of 8.5 million litres of jetfuel per year. However the real sustainable jetfuel production would certainly be less.

Firstly, oil palm residues represent the overwhelming quantity of the theoretical maximum feedstock. But only 20% of the area of productive oil palm in Malaysia today, is certified under RSPO, the only internationally recognised scheme of sustainable oil palm plantations. The sustainable feedstock from oil palm is 5 times smaller than the theoretical maximum. Taking in account only 20% of sustainable oil palm residues, Malaysia can produce a maximum of 3.8 million litres of sustainable jetfuel per year.

Secondly, given widespread international environmental concerns on any kind of oil palm, it could even be possible that some international airlines flying to and from Malaysia would prefer avoiding any blend from oil palm residues. Rubberwood, sugarcane, coconut and paddy cultivations are not linked to deforestation, and their lignocellulosic residues probably wouldn't create any major environmental concern among the international community. These crops are extremely important for the livelihoods of Malaysian smallholders. The entire forest production of Peninsular Malaysia is eco-certified under PEFC, the world's largest forest certification scheme. Thus, excluding palm oil and sourcing only on other sustainable feedstocks, Malaysia would be able to produce up to 2.5 million litres of sustainable jetfuel per year.

Entire Malaysia consumes over 3 billion litres of jetfuel per year. Of all passenger traffic, approximately 5% travels to Europe. Consequently, it is possible to achieve up to 6% of biofuel blend on all flights to Europe, departing from Malaysia. Similarly, it is possible to achieve up to 3% of sustainable biofuel blend on all flights to Europe, or up to 2% of sustainable biofuel blend if excluding oil palm feedstocks.

10
LCA of agricultural production systems – the case of paddy

Environmental Life Cycle Assessment (LCA)

A comprehensive literature review was conducted for the main agricultural crop production systems in Malaysia. The goal of this review was to identify relevant research, available in the international literature in relation to the environmental aspects. As described in the previous chapter, the overall environmental assessment is of interest at this stage, thus the review was not limited to Life Cycle Assessments of the crops but it also included other environmental frameworks such as Greenhouse Gas Accounting, Energy and Emergy Analysis, and Environmental Impact Assessment (EIA). An additional reason to consider other environmental frameworks was the limited, in most cases, Life Cycle Assessment studies conducted for the Crop production systems in question. Finally, in certain cases, a narrow literature review was performed on the processing and production of biofuel from the products and by-products of the crop production systems in question.

The analysis of biomass availability, logistics costs and possibilities for biomass collection, biomass conversion processes, and socio-political context for biofuel promotion, will allow defining bio-jetfuel development scenarios in Malaysia. Environmental Life Cycle Assessment (LCA) will then be used to predict the potential environmental consequences, from well to wheel, of the implementation of these scenarios and to help decision making. This chapter aims at defining the methodological framework for the environmental assessments to be carried out, addressing the main issues and requisites.

Since the last twenty years saw an important increase in environmental LCA studies applied to food and bioenergy production systems and a rise in methodological issues specific to biomass-based systems, the content of this chapter is mainly based on a literature review about these items, following the classical LCA pattern (see Figure 28): Goal and scope definition, Life Cycle Inventory (LCI), Life Cycle Impact Assessment, and Interpretation.

The case of Paddy

The analysis of the studies, focusing on the environmental impact of various agricultural practices, lead to a number of practical conclusions:

• Most of the paddy rice experimental studies are more interested in the comparison of the impact of different agricultural practices than the simple calculation of the Global Warming Potential of the product.

• Paradoxically, Global Warming Potential seemed to decrease with the increasing rate of application of inorganic fertilizers. A similar rate of increase in yield of paddy rice was not identified.

• According to the studies, the normally practiced fertilization rates, between 100-200 kg N/ha, in most farms have little or no effect on field emissions of paddy rice. Higher fertilization rates have to be implemented in order to see a visible decrease of methane (CH_4) emissions.

• Organic amendments increased Global Warming Potential and seemed to be generally worse in yield-scaled Global Warming Potential than the standard NPK fertilizer treatments.

• Biomass residue incorporation increased CH_4 emissions by a factor of 2-10, across completely different system configurations. This casts a shadow to the sustainable character of incorporation strategies.

• Biogas residues or composted manure, on the other hand, seemed to have little or no effect on CH_4 emissions. This is probably due to the lower C/N ratio of the composted or digested biomass. Rice straw has a quite high C/N ratio (around 60).

• Irrigation practices that introduce drainage events seem to have a considerable effect on CH_4 emissions. They also generally seem to increase carbon stocks and yields of fields.

• No tilling was shown to have lower CH_4 emissions than conventional tilling techniques. A part of this reduction comes from the reduced mechanical work required and the avoided combustion of fuels and use of electricity.

• A correlation was identified between organic carbon in the soil and reduced CH_4 emissions. Perhaps strategies that induce increase of carbon stocks in rice field should be aimed, in order to reduce their Global Warming Potential impacts at the same time.

Synthesis

The choice of a broad environmental assessment shows its strengths in the case of rice. Of course data availability and consistency of methods play a role. But the way the methods are connected that leads to a better understanding of the crop's peculiarities is clear. The Life Cycle Assessment studies of the crop, as expected, suggested a variety of different systems, with many different assumptions and methods, based on different cropping systems, in different areas of the world. However, it was precisely all these

differences that allowed us to make the necessary step: to identify the main environmental hotspot of the crop and the culprit responsible for creating it.
It was clear, after understanding that Climate Change was the most interesting impact category for rice, and the investigation on the main determinants of Greenhouse Gas emissions, that field emissions and specifically those of CH_4 were responsible. The studies that were investigated afterwards were selected according to this criterion. However it was interesting to discover that these studies, specifically focusing on Global Warming Potential emissions of paddy rice cultivation, were not really interested in the simple quantification of Global Warming Potential emissions but rather in the understanding of the impacts different agricultural practices and mitigation strategies had to these emissions.
The study then naturally focused on identifying practices that reduce CH_4 field emissions, and so reduce the total Global Warming Potential of paddy rice production. But still, these insights are only valuable when they are connected to the context of a 2nd generation biokerosene production system, the focus of this study.
Opportunities and barriers for the use of biomass residues, the goal of the research, can be identified from the previous analysis. When it comes to biomass residue management, it was clear that their incorporation to the fields, a mitigation strategy normally seen as sustainable, lead to an increase of CH_4 emissions and total Global Warming Potential of paddy rice cultivation.

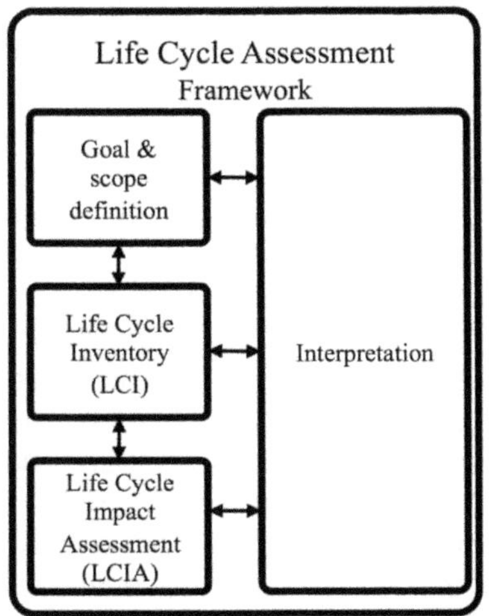

Impact assessment category	Number of studies
Cumulative energy demand	2
Water use	2
Eco toxicity	2
Ozone depletion potential	2
Resource demand	2
Human toxicity	4
Photochemical oxidant formation	4
Eutrophication potential	8
Acidification potential	8
Global warming potential	10

Figure 28: Approaches and environmental assessments results that are relevant to Malaysia.

Since the cultivation phase was previously identified as the major contributor in every bioenergy supply chain, the decision not to incorporate these residues in the field provides both a decrease of Global Warming Potential and an opportunity for feedstock appropriation. The paddy rice growers, by selecting to remove the biomass residues, that is rice straws, from the field can reduce their environmental impacts, create a new revenue from these resources, and provide a valuable input to the biokerosene production system. Furthermore, it was discovered that residues that have been previously processed, either by anaerobic digestion or enzymatic hydrolysis, have their C/N ratio lowered and can be transformed into biomass residues that can be incorporated into the fields without increasing CH_4 emissions. Therefore, increasing returns can be achieved by the connection between paddy rice production and biokerosene production, benefiting both parties.

The central problem of biomass residue extraction, that is the eventual reduction of the yield and carbon stocks in the field, can also be mitigated by other strategies that also reduce total Global Warming Potential of cultivation. A strong correlation was found between increasing fertilizer inputs and decreasing CH_4 emissions, while providing the expected results of increasing the yields of paddy rice cultivation. Therefore, more rice and more biomass residues can be produced, residues that bring along a lower Global Warming Potential impact. This strategy therefore is beneficial to the short-term performance and long-term sustainability of a biokerosene production system. Reducing the Global Warming Potential of cultivation and ultimately providing biokerosene that is even more advantageous against fossil kerosene improves the short-term performance of the system. Providing consistent, or increasing, biomass production in the region, despite the extraction of residues, assures the long-term sustainability of the system. In the same way, irrigation practices that allow for a partial drainage of the fields were shown to also decrease the CH_4 field emissions significantly and increase yields, so they are also seen as beneficial to the farmers and the BPS.

It is clear, from this analysis, that mitigation strategies aimed at reducing environmental impacts of cropping systems, in this case paddy production systems, are not in conflict with strategies that aim towards an implementation and a sustainable development of a 2nd generation biokerosene production system. More research is required on the above strategies, the share they represent in the typical strategies of Malaysian growers, and the difficulties of transitioning from other business-as-usual strategies to them. All these issues should be discussed with the growers and producers of paddy rice and will have a great impact on the feasibility and limits of using paddy rice biomass residues.

11
Logistics and transportation

Logistic potentials and constraints

Biomass supply chain and logistics issues

Malaysia covers 33 million ha, among which the peninsular represents only 13 million ha. The topography is very variable and has strong influence both on the land use and on human activities. Historically and still today, most of the agricultural development, industries and transport network spread on the West side of the central mountain range that runs along the peninsula. Transport and general logistic costs make the larger share of the components of supply cost for most of the biomass for energy (Giampietro et al., 2006; Athanasios et al., 2008). The fragmentation of the resource tends to increase collection and supply costs (Sturmer et al., 2011) despite a rather good transportation network (Rosena, 2008). Malaysian industry players have identified generic difficulties applying to the Peninsular Malaysia, especially issues regarding:

- integration and coordination,
- management techniques among the supply chain companies,
- with Information technology / electronic data interchange,
- skilled and trained manpower,
- sources of logistics data and information and dissemination of it,
- assistance to local service providers (LSP),
- research and development (R & D) of the industry,
- regulatory tools for facilitation.

Sabah and Sarawak have a limited road network, with a lot of connections made through tracks or river transport. Their location across the South China Sea hampers their connectivity to transport low value biomass, while their higher value goods such as timber products and palm oil often travel through Chinese harbours or other neighbouring countries, rather than through Peninsular Malaysia. Conversely, most of the peninsula west coast is well equipped with a dense road network and highways and one Singapore to Thailand railway. The east side of the peninsula, which offers most of the biomass standing stocks (plantations and forests) is much less connected by

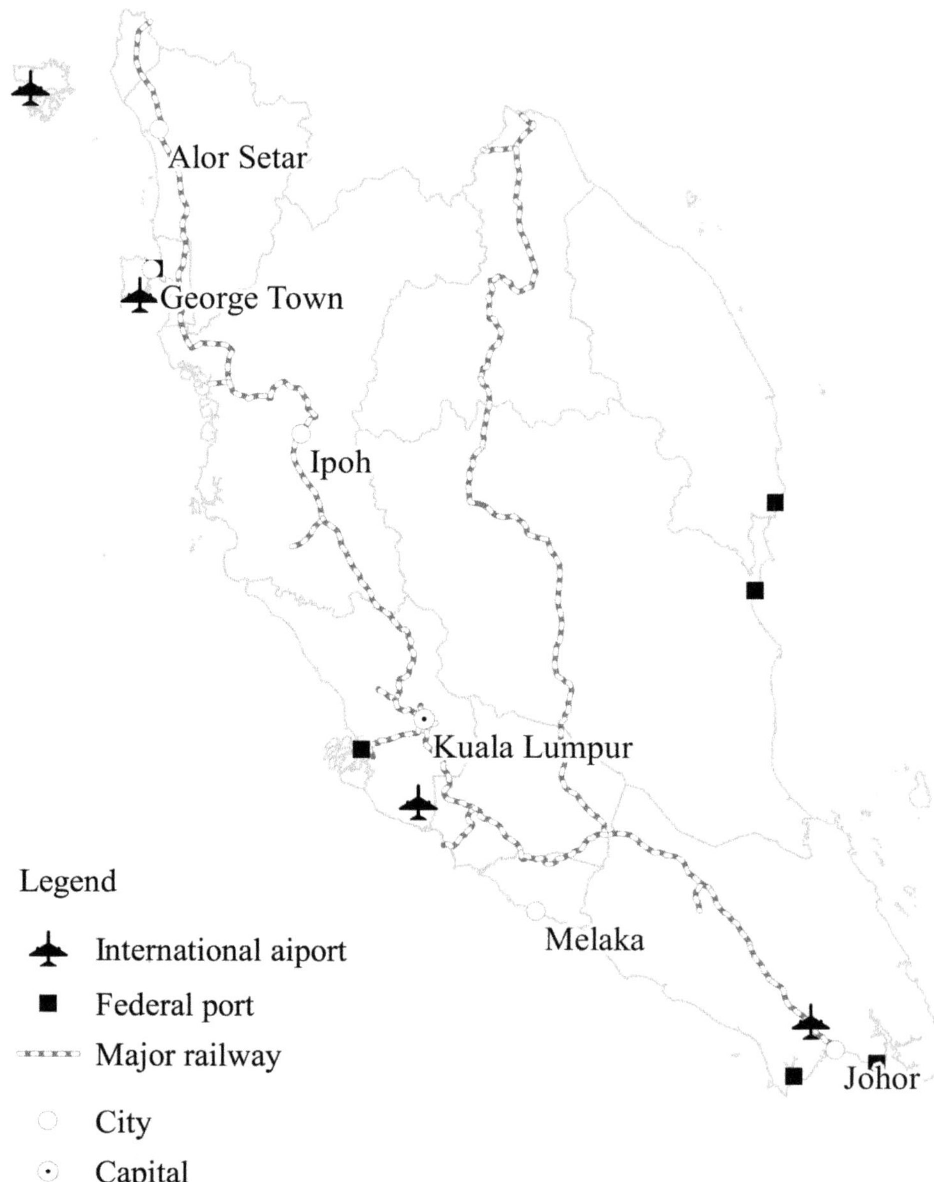

Figure 29: Location of major logistic connections in Peninsular Malaysia.

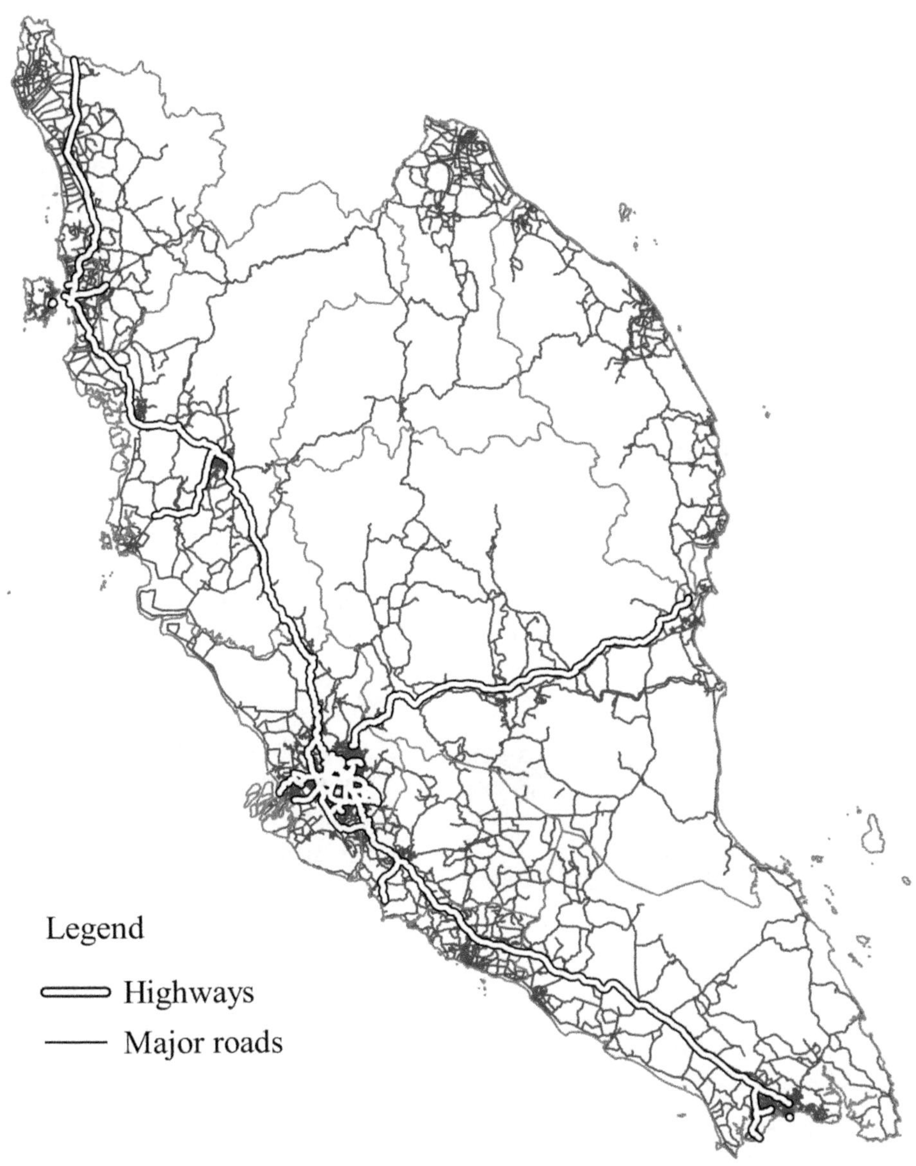

Figure 30: Road network in Peninsular Malaysia.

the road with only one easy West-East corridor. It connects the two coasts through a lower passage in the central range. Other lesser roads connect the East coast but imply long transportation delays.

One major international airport hub is located near Kuala Lumpur, and 3 other airports are classified as "international"; Johor, Penang, and Langkawi) but attract much less traffic than Kuala Lumpur and Singapore. Singapore, just across a bridge from Johor is a major hub of Southeast Asia, and could easily represent another market for bio-jetfuel sourced from the peninsular biomass.

There are six federal ports in Peninsular Malaysia. The East coast ports chandelle oil and gas products, and could represent an alternative to the limited East-West road connections, provided that some multimodal solutions could be economically developed for biomass or for bio-jetfuel, according to handling constraints of raw material or according to the location of the potential refineries.

Biomass supply chain and logistics issues

The cost of raw material depends on the distances, and on the size of the trucks. There are 4 common categories of trucks in Malaysian logistics. The biggest, 26T "prime mover" is never seen transporting raw material, and is adequate for finished products. 10-12T trucks can be seen supplying the big mills, which have the capacity and the purchasing power to secure large

1-2T

3T

Prime mover - 26T

10-12T

Figure 31: Examples of the 4 trucks categories: 1T, 3T, 10T, and 26T trucks (source: this study, web pictures).

harvesting auctions. Smaller trucks, especially the 1T trucks, are the most common to source from the scattered resources.

Each of these truck categories implies a different cost curve, the smaller truck implying the higher transport cost in Ringgit Malaysia per tonne per kilometre.

The following scenario shows the dramatic influence of truck size on the cost of raw material: if a company contracts a transporter to carry rubber logs over 150 km, the prices varies dramatically from more than 3RM/T/km with a 1T truck, to around 1RM/T/km with a 10T truck. A similar scenario illustrates the effect of the combination of the transport cost factor versus the location of a company and the uneven spread of the resource. Given a profitability of the transport being limited to 150 RM/T only, companies situated in Muar (in Johor) or in Sungai Buloh (in Selangor), could have different access to the resource, by using different size trucks. Hiring 1T trucks would make more than 97% of a given resource, excluded from the reach of a biorefinery in Johor. However, it would exclude only 93% of it if the refinery is in Selangor. Conversely, hiring 3T trucks would make the first company able to access 14% of the resource, the others would access up to 30% of it.

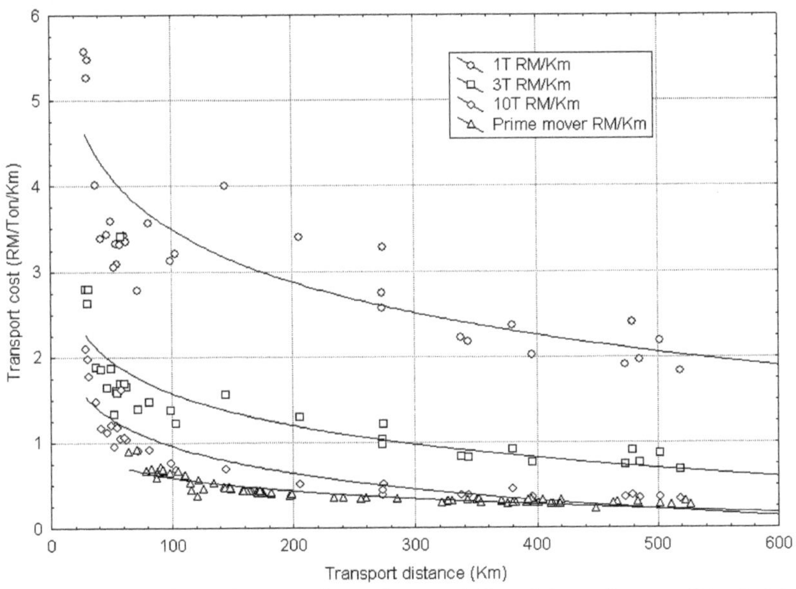

Figure 32: Cost curves of 4 truck categories vs transportation distance (source: this study).

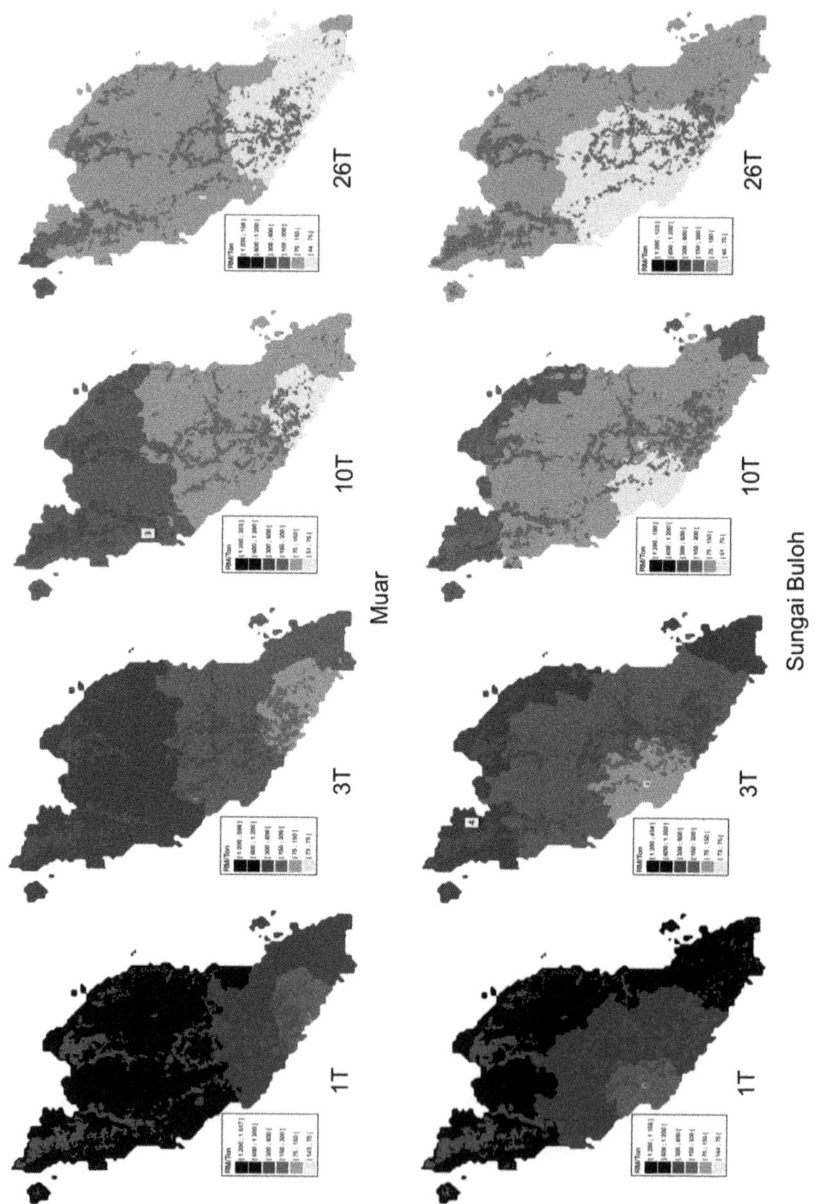

Figure 33: Shares of rubberwood resource economically accessible to: respectively Johor and Selangor possible biorefineries, when transporting logs with 1T, 3T, 10T, and 26T trucks, for a profitability threshold of RM150/T (source: this study).

Vehicle	Distance	RM/tonne/km
1 tonne truck	150 km	RM3++
3 tonne truck	150 km	<RM2
10 tonne truck	150 km	<RM1

Table 16: Transportation costs over a 150 km distance for different truck sizes (source: this study).

Vehicle	% if the resource excluded from Sungai Buloh	% if the resource excluded from Muar
1 tonne truck	93-96% excluded	97-99% excluded
3 tonne truck	70-73% excluded	86-89% excluded
10 tonne truck	27-30% excluded	33-36% excluded
26 tonne truck	0% excluded	10-13% excluded

Table 17: Percentage of the resource excluded, with a transport profitability threshold of 150 RM/T, for different truck sizes, and for 2 demand centres (source: this study).

Procurement consolidation

The major drawback of the resource supply is the sensitivity of supply cost to the uneven spread of the resource. This makes the location of biorefineries critical for the long-term feasibility of biofuel production. This effect is the main reason for the potential difficulties faced by some biorefinery locations for a given resource, as per the above scenarios.

However, tackling the problem upstream, at the transport stage, would have a huge lever effect on the situation. For example, the panel industries have a very high purchasing power because of the big size of their industrial units. This makes them able to buy large quantities and to consolidate transportation with big trucks. As their purchasing power is proportional to their capacity, and also to transport distance, one can compute the limits of the zones where each of these industries is virtually the strongest purchaser.

In order to emulate them, the possible biorefineries could consider "group buying" and consolidated logistics. As shown with the example of rubberwood transportation, a supply consolidation, which would allow using bigger trucks, makes easily 70 to 100% of the resource economical to access by the smallest players, just by the scales economies gained using bigger trucks.

12
Socio-political framework

Trends: the biofuels are seen as controversial

The main negative concerns about biofuels are:
- competition with food,
- doubts about their ability to meet some or all of their stated objectives.

The global production of biofuel is currently estimated around 100 billion litres (HLPE, 2013). It raised issues about competition with food and arable lands (in the recent years, grain and edible oil prices increased by 70 to 120%, world food markets experienced the largest price shock in 30 years).

Competition can happen through direct land use change (on arable lands, direct replacement of food crops by energy crops), thus decreasing the production of food. It can be through indirect land use change. In that case biofuel crops may change market conditions, either resulting in crops and livestock areas to be less profitable, or in diverting existing productions from food use to energy use. Today's biofuel production probably mobilizes around 2 to 3% of arable lands globally (HLPE, 2013).

Issues raised concerning biofuels also include climate change, energy security, and declining oil reserves. For example, some biofuel production systems actually emit more greenhouse gases than fossil fuels. Others might be sustainable, but unable to produce the vast quantities of energy required.

Advanced biofuels are emerging: new policies try to shift towards second generation biofuels which are produced from residues of crop, forest, and industrial wastes (non-food). Few tropical countries have the resources to move forward to second generation biofuels, given the often proprietary nature of this technology, the high capital investments required, and the induced requirements for infrastructures, logistics and human capital (HLPE, 2013).

The international regulatory framework

As a result of the criticism toward biofuels (especially the 1st generation), several governance mechanisms have emerged: they may take the form of legislation, international agreements, jurisdictional guidelines, company policies or market-based certification schemes. There are several recognised certification schemes, following two main models: mandatory certification or voluntary certification. Note that NGOs like WWF, with the help of the private sector, developed private certification schemes to overcome the lack of national regulations concerning biofuel production and certification.

Mandatory certification for feedstock and biofuel sustainability

The most famous mandatory certifications achieving sustainability objectives are the Energy Independence and Security Act (EISA) from the USA, and the European Union's Renewable Energy Directive (EU-RED).
The EISA stipulates standards for biofuels, with required GHG emissions profiles compared to conventional petroleum fuels emissions.
There are three ways of complying with the EU-RED:

- To prove evidence of compliance with the national member state system where the biofuel is used,
- To meet the terms of a bilateral or multilateral agreement approved by the European commission (EC),
- To refer to a voluntary certification scheme approved by the EC.

In Asia, the main example is the Indonesian Sustainable Palm Oil. It is mandatory for all palm growers in Indonesia, and stipulates standards for agricultural practices.

Voluntary certification for feedstock and biofuel sustainability

This kind of scheme is linked to voluntary adoption of production standards and certifications evaluated by an independent third party: the producer has to cover costs for complying with the standard, or has to transfer this cost down to its customers. Three certification schemes can address biofuel feedstock production, processing and trade:

- the Roundtable on Sustainable Palm Oil (RSPO),
- the Roundtable on Sustainable Biomaterials (RSB),
- the International Sustainability and Carbon Certification (ISCC).

Three more certification schemes are not strictly related to biofuel feedstock production, processing and trade. They certify the performance of forestry operation. However, these certifications are very important for bio-jetfuel because they are relevant to second-generation biofuel projects using the woody biomass:

- the Forest Stewardship Council certification (FSC),
- the Malaysian Timber Certification Scheme (MTCS),
- the Programme for the Endorsement of Forest Certification (PEFC).

They still lack aspects on GHG emission reduction, but somehow incorporate LUC issues since they ban deforestation. The FSC scheme may gain future relevance with second-generation biofuel development because it is the only certification scheme, which prohibits the use of genetically modified organisms.

Roundtable on Sustainable Palm Oil

The RSPO is the major certification in Malaysia and probably the most visible internationally because of the controversies around palm oil. The RSPO, established in 2004, aims to include all stakeholders, from producers and consumers to banks and environmental conservation groups. It essentially focuses on agricultural or feedstock production aspects (does not cover transport and processing). The RSPO Principles & Criteria consist of 8 principles and 39 criteria with 123 specific national indicators: they form the performance indicators for RSPO Certification. However, a lot of criticisms have been arised on its efficiency in addressing environmental concerns (e.g., the role of the RSPO in protecting rainforests).

Roundtable on Sustainable Biomaterials

The RSB was created in 2007, and it applies to all operators and products in the biomass and biofuels industry. It is a voluntary standard proving that your product is responsibly produced. The European Comission recognizes the RSB certification: it allows access to the EU market as sustainable biofuels. It is the only standard which covers the entire value chain from farm to end-user. The certification requires: 1) proven 50 % cut in GHG emissions for a blend of biofuels, compared to fossil fuels, 2) the operators have to meet regulatory GHG requirements in the region where they operate, 3) compliance with 12 principles and criteria. However, there are some criticisms made today on this certification in the form of the lack of local or regional adaptations of the scheme, and on the laws themselves. In terms of environmental and social criteria, WWF denounces its weak spots with respect to the criteria for handling non-GMO materials. Furthermore, the RSB does not include specific direct or indirect land use changes. In Malaysia, this certification is not well known, even for the stakeholders: it seems to them that it is more expensive and harder to get, than ISCC.

International Sustainability and Carbon Certification

The ISCC focuses on greenhouse gas reduction through the value chain, sustainable land use, protection of natural habitats and social sustainability for the feedstock production. This certification can be both an alternative and a complement to the RSB. Based on the EU-RED, ISCC requires a minimum GHG emissions saving of 35% (rising to 50% in 2017 and to 60% in 2018; that is to say for installations in which production starts from 2017 and onwards). Feedstock production also needs to comply with 6 principles. Most of the Malaysian biofuel companies are certified ISCC. If a company or a plantation is RSPO certified, it can easily switch to ISCC (it is much harder to switch from RSPO to RSB: contrarily to the RSB, in RSPO there are no specific criteria on the residues, wastes and by-products). The main criticism on this certification is that a social and environmental management system is not explicitly required, for example there are no requirements in connection with the spraying of pesticides or health protection, working hours, and remuneration of the workers.

Forest Stewardship Council certification

The FSC (founded in 1993) is an independent non-governmental and non-profit organisation to promote sustainable forest management. The framework includes 10 principles and 56 criteria. Two of the major criticisms of this certification are that it is often impossible to obtain for small and medium sized enterprises in developing countries, due to the high cost involved, and that it has not really contributed to curb deforestation since its creation.

Malaysian Timber Certification Scheme & PEFC

The MTCS is developed under the Programme for the Endorsement of Forest Certification (PEFC). The Malaysian Timber Certification Council was established in 1998 as an independent organisation to develop and operate the MTCS. As a voluntary national scheme, the MTCS provides for independent assessment of forest management practices, to ensure the sustainable management of Malaysia's natural forest and forest plantations, as well as to meet the demand for certified timber products. The MTCS is the first tropical timber certification scheme in the Asia Pacific region to be endorsed by the PEFC, the world's largest forest certification programme, representing more than 200 million ha MTof certified forests worldwide. The entire Peninsular Malaysia productive forests have been certified under this standard for the management, which represents a total area of 4.7 million ha, or 26.6% of the total forested area in Malaysia.

Situation of feedstocks in Malaysia

Malaysia's main agricultural crops are oil palm, rubber trees, cocoa, paddy and coconut. Malaysia, like Indonesia, has large palm oil plantations. The two countries account for more than 80% of total palm oil production in the world (Mukherjee et al., 2014).
Except paddy, all other crops have experienced a decline in their cultivated area. Oil palm plantations, more profitable for both for smallholders and for big companies, gradually replace the other crops.
Approximately 20% of Malaysian oil palm productive plantations are certified under the RSPO scheme.

Malaysian agriculture: a development issue

Malaysian agriculture has drastically changed since the independence of the country. Figures from 1957 show that the primary sector accounted then for 46% of GDP against 14.3% in 2003. The country has undertaken several development policies in the rural sector since independence. At that time, the country had to face 3 major problems:

- rural poverty affecting mostly the Malay rice farmers,
- difficult governance of the country, due to the communist guerrillas,
- low productivity of the agricultural sector causing a serious rice deficiency.

National agricultural policies have subsequently focused on land use conversion from forest to agriculture, a strategy to address the problem of landless agricultural households, and to increase the income of rural communities. The smallholders sector is operated by individual farmers. The collective acreage of land operated by smallholders now amounts to 75% of the total area under agriculture. They are the main contributors to food crop production as well as industrial crop production (oil palm, rubber, cocoa, pineapple) with the help of development and management federal organisations (FELDA, FELGRA, RISDA etc.). The typical smallholdings range from 2 to 4 hectares. Conversely, the estate sector is characterised by holdings of more than 40.5 hectares, generally owned by private companies and public-listed corporations. It has progresively decreased, now representing 25% of the agricultural area.

Forest management and governance

National parks and wildlife reserves are managed by specific institutions such as the Forestry Department Peninsular Malaysia (FDPM) at federal level, and Forestry Departments in each state. As their function is dedicated to conservation, these areas are not considered for this biomass survey.

Each State Forestry Department is divided into Forest Operations and Forest Development Divisions. These Divisions are supported by the District Forest Offices. The National Forestry Council, created in 1971, is the organisation in charge of facilitating the coordinated work between the 2 levels of power in order to plan the forest management and maintain the forest as a long term renewable resource. The decisions of the Council depend on the state governments except if the matter falls under the federal government jurisdiction. The forestry laws concerning management, planning and forest renewal operations were standardized and strengthened in 1978 under the National Forestry Policy. In 1984, this Policy was revised through the National Forestry Act in order to better deal with environmental protection, the conservation of biological diversity, forest encroachment and illegal logging.
The National Forestry Policy was amended in 1993 to strengthen some aspects of the policy (after Rio Conference Rio, 1992). These amendments were welcomed because they suited with the nation's aspiration in pursuing a better living quality for the future generations. Even if Sabah and Sarawak developed their forest policy independently, they share many similarities with Peninsular Malaysia, thanks to the implementation of the National Forest Council whose role is also to help in adopting common approach to forestry issues.

Feedstocks perspectives

Forest sector

Forest and wood residues offer a significant potential of biomass. But the use of residues directly produced in the forests, is often subject to criticism from public opinion, being linked to logging activities. Furthermore, removing more biomass, detritus and dead organic matter from the forest floor might harm the regeneration function of forests as it causes depletion of the soil's fertility. Tropical soils are highly dependent on dead leaves and decomposing plant material to maintain nutrient and carbon cycles. This issue could lead to conflicts with NGOs, and compromise the current state of certifications.
Using wood/tree-based by-products and wastes as a potential source of biomass, offers synergy opportunities. Today, Malaysia is the second largest tropical timber exporter in the world. The timber industry is widely developed and contributes to various sectors such as furniture and components, panel products, mouldings and joinery, and construction. Scraps, split logs, sawdust and otherwise factory floor wastes at the processing and manufacturing level would be excellent feedstocks, rather than to be burned in open air for their disposal. Note that the idea of using process wastes is not new, many by-

products already have an important commercial use. The problem concerning these products is the wastes availability with some industry players who have already developed cogeneration systems, and already use all their wastes. However in the Malaysian industry, the complete use of wood residues for cogeneration is far to be widely developed, and most it still forms a large source of available biomass.

Paddy

Since the Third Malaysian Plan (1976-1980), Malaysian policies have supported the rice sector. With the 2008 world food crisis, the Malaysian government considers food security as an integral national policy, which is synonymous of rice security. In 2012, Malaysia produced 62.5% of its domestic consumption but the aim is to be 100% self sufficient by 2015. Paddy is the only crop that has not experienced a decrease in its area cultivated with the oil palm. Seen from a political-economy perspective, paddy seems to be one of the most suitable crops for bio-jetfuel, because a balanced use of its residues would not cause any major environmental problem. There are some challenges concerning the global sustainability of the sector. For example, some farmers stop their activities for better economic alternatives such as jobs in the industrial sector. Moreover, the government has launched programmes, especially under Bio-TechCorp, to develop GMO paddy in order to reach rice self-sufficiency. Even if using GMO products is still accepted in RSB or ISCC certifications, it is still a potential source of controversies for European end-users.

Sugarcane

Lignocellulosic wastes from sugarcane contain high level of energy that could be an important potential for bio-jetfuel, but there is neither specific public policy nor incentive on this crop. The 2010 National Agricultural Policy did not give attention to increase or improve sugarcane production compared to oil palm or other crops. The production of sugarcane in Malaysia is small; the country depends on sugar import. From an environmental point of view, sugarcane is a water-intensive crop, which can create problems in regions with dry seasons, but this is not an issue in the Malaysia because of its important and even rainfall.

Rubber

For a long time rubber was a strategic sector in Malaysia but the national economic structure was changed since independence through diversification of export products, and this sector has been critically reduced. The Malaysian Government has taken steps to provide better incentives through the Pioneer Status and Investment Tax Allowance schemes (Investment Act, 1986). Despite those 2 initiatives, rubber plantations remain financially difficult for small farmers. Even if some of the policies and programmes (2010 National Agriculture Policy, ETP) include the support of the sector, the discourse hold by the government through the Ministry of Plantation Industries and Commodities do not consider it as a promising sector for biofuel production. Almost all of its lignocellulosic residues are already valorised for furniture making of for the board and panels industry. The supply costs of rubberwood are pushed up by this demand. Thus any biofuel valorisation would be economically difficult.

Coconut

This crop is not suffering major criticisms and controversies around the world, neither from an environmental point of view, or from a social perspective. There is currently no specific programme or initiative for this sector. A national conference was organized in 2009 to “revitalize the coconut industry” but it did not end up with promising solutions. Given its limited quantities of available residues, this crop would not be a major feedstock, but it can be used as a perfect complementary source of biofuel.

Oil palm

This crop is the most supported in Malaysia (National Biofuel Policy/ Economic Transformation Programme/ National Biomass Strategy, etc.). The focus on energy seems to favour biodiesel for land transportation, using palm oil. But the palm oil market price is so high that no biodiesel plant can be profitable without government subsidies. So far, lignocellulosic residues offer huge quantities, and could be a very large feedstock for bio-jetfuel. However, there is already a developing trend for the use of these residues: furniture makers are already starting to valorise it for composite wood, and cattle industry is beginning to feed animals with the stems, for their high contents in sugars. These emerging markets could develop quickly and end up in making market prices sufficiently high for a difficult economic competition with any bioenergy valorisation, as it already happened with rubberwood. Nevertheless, the immediate risk lie in the possible environmental criticisms by the international community, for the direct and indirect links of oil palm with deforestation. Only 20% of Malaysian oil palm productive areas are certified as sustainable under RSPO, and RSPO itself is criticised by some NGOs.

CROPS	Oil Palm	Forest	Paddy	Rubber	Sugar cane	Coconut
Climate Mitigation	-	✔	✔	-	-	-
Poverty Alleviation	✔	✔	✔	✔	✔	✔
Governement Incentives	✔✔✔	-	✔	✔✔	-	-
Residues Availability	✔✔✔	✔✔✔	✔✔	-	-	-
Main product Availability	-	✔	-	-	-	-
Land Extension	-	-	-	-	-	-
Land Area	✔✔✔	✔✔✔	✔	✔	✔	✔
International opinion	✖✖✖	-	-	-	-	-
Competition with food	✖	-	-	✖	✖	-
Deforestation potential	✖✖✖	-	-	-	-	-
Logistics difficulties	✖✖	✖✖	✖	✖✖✖	✖✖✖	✖✖✖

Figure 34: Opportunities and risks according to feedstocks.

Situation of biofuels in Malaysia

Through the National Biofuel Policy (2006), the Malaysian Government has introduced the use of B5 blended biodiesel (95% petroleum diesel and 5% biofuels). In 2006 and 2007, 92 biodiesel projects have been approved in Malaysia. In 2012 most of them have been closed because of high international prices for crude palm oil. It makes most of the biodiesel plants difficultly profitable without heavy subsidies. In 2013, the Malaysian government announced that B10 blended biodiesel would become mandatory to encourage again the biodiesel industry.

In Malaysia the biofuel sector is slowly emerging. A national objective is visible; through the "2020 biomass strategy". For now the government attention mostly focuses on the oil palm sector.

The creation of the Centre of Excellence on Biomass Valorisation, with collaboration between the industrial sector, universities, research organizations and governments has become one of the important tools to reach a real national development.

National policy on biotechnology

To support biotechnology as one of the key strategic drivers to develop the country, the Ministry of Science, Technology and Innovation implemented the National Biotechnology Policy and Biotech Corp (a dedicated biotechnology agency). The policy covers the biofuel sector and the aviation sector.

Inventory of past and current biofuel incentives

Due to the energy crisis in 1973, the federal government has implemented policies to improve energy security (Petroleum Development Act, National Energy Policy, National Depletion Policy) and to diversify the sources only based on petroleum and gas (Four Fuel Diversification Policy). Malaysia started developing these policies also to promote employment; raise incomes and boost export earnings.

Although, the government biofuel policy assumes that biofuels will reduce greenhouse gases emissions and pollutions, the environmental issue was at first not a priority. The first policy on renewable energy was introduced in 2001 to reduce dependency on fossil fuels.

In 2006, the government developed ambitious biofuel policies through the National Biofuel Policy. The policy provides the overarching framework to develop biofuels as one of the five main energy sources for Malaysia. The main point was B5 blended biodiesel launched in phases, starting with the central region.

In 2007, the Biofuel Industrial Act was enacted with the purpose of providing legislation for the mandatory use of biofuel, licensing of activities relating to biofuel and for matters connected therewith and incidental thereto. The Ministry of International Trade and Industry established a licensing system (the Act was enforced on 2008).

The National Green Technology Policy was implemented in 2009 in order to reduce GHG emissions, use renewable energy resources, and use natural resources with one of the four pillars in the transportation sector.

There are 2 other policies to take into account in the Malaysian biofuel development: the 2010 National Policy on climate change and the 2010 National Agricultural Plan.

In 2013, the Malaysian government announced the implementation of a mandatory B10 blended biodiesel. It aims to be consistent with the United Nations Framework Convention on Climate Change (UNFCCC), but only on the assumption that, by definition, biofuels would help to reduce greenhouse gases emissions. There are no specific criteria to assess and ensure that it is effectively the case, for a given implementation.

Future biofuel objectives

Several policies based on biofuel strategies have been planned: the third master industrial plan (2009-2020), and the National Timber Industry Policy. These planned policies focus mainly on the use of oil palm biomass and this trend is confirmed through the National Biomass Strategy 2020.

The main incentives are:

- the Bionexus Status,
- the Pioneer Status,
- the Investment Tax Allowance,
- the Economic Transformation Program (ETP).

With the Bionexus Status, companies participating in value-added biotechnology and/or life science activities can receive tax breaks and apply for funding.

The Malaysian Investment Development Authority (MIDA) awards the Pioneer Status to manufacturing businesses. The eligible companies receive 30% exemption from taxable statutory income for 5 years and 100% exemption if investments are made in promoted areas (Sabah, Sarawak, Perlis and designated “Eastern Corridor” of Peninsular Malaysia). In the near future, 100% exemption will also be provided to “high value added” production in non-promoted areas.

The Investment Tax Allowance is awarded by MIDA to companies with high value added. Companies are allowed to offset 60% of the qualifying capital expenditure they incur within five years against 70% of their statutory income. The allowance increases to 100% if the investment is made in a promoted area. Going forward, 100% exemption will also be provided to "high value added" production in non-promoted areas.

In order to elevate the country to developed nation status by 2020, the Government created the Economic Transformation Program (ETP): 12 identified keys areas have been defined. Biofuels under the Key area "Palm oil & Rubber", by developing biogas facilities at palm oil mills, and commercialising second generation biofuel using non-food oil palm biomass.

Malaysia aviation and aerospace sectors

Since 2012, there are discussions about launching a National Aviation Policy to ensure free and fair competition on a level playing field between the air transport division of the Transport Ministry and relevant stakeholders in the aviation industry (Malaysia Airlines, Air Asia, Malindo and Malaysia Airports Holdings Bhd). However, in 2014 there is still no established policy for the aviation sector, the main discussions are still about strengthening the services network in the aviation industry, upgrading airports and creating human capital.

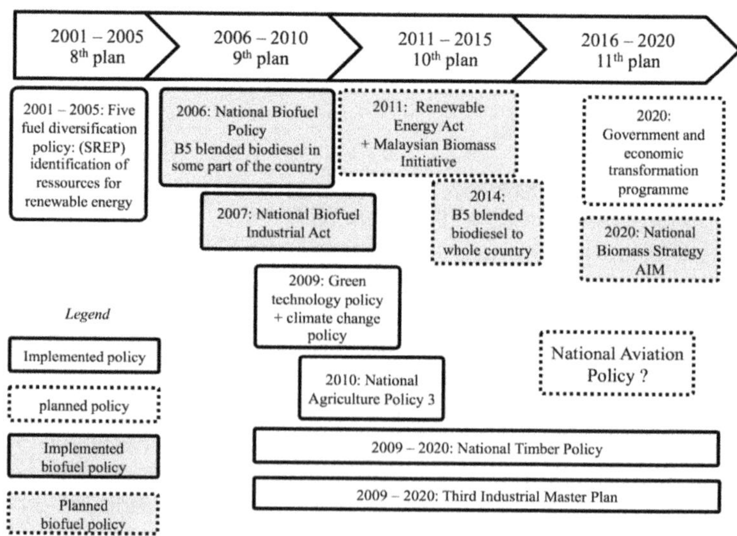

Figure 35: Malaysia incentives and plans linked to biomass and bio-jetfuel.

The National Aerospace Blueprint (launched in 1997) encourages the aerospace industry to become a growth sector. In 2001, the Malaysian Aerospace Council was created: it approved initiatives such as Aerospace Malaysia Innovation Centre (AMIC, created in 2011). It is a consortium between the Malaysian government through Malaysia Industry Government Group for High Technology (MIGHT) and the industry through Airbus Group. It coordinates R&D activities from universities, research institutes and companies (UPM, CIRAD, etc.). It is the main promoter of the use of alternative jetfuel for aviation in Malaysia.

Among other programmes are:

- the National UAV Programme,
- the Maintenance Repair and Overhaul (MRO) plan,
- the System Capability Development Programme,
- the Satellite Development Programme,
- the Leader Aerospace Programme,
- the Next Generation Aircraft Programme,
- the Composites Technology Research Malaysia Sdn Bhd.

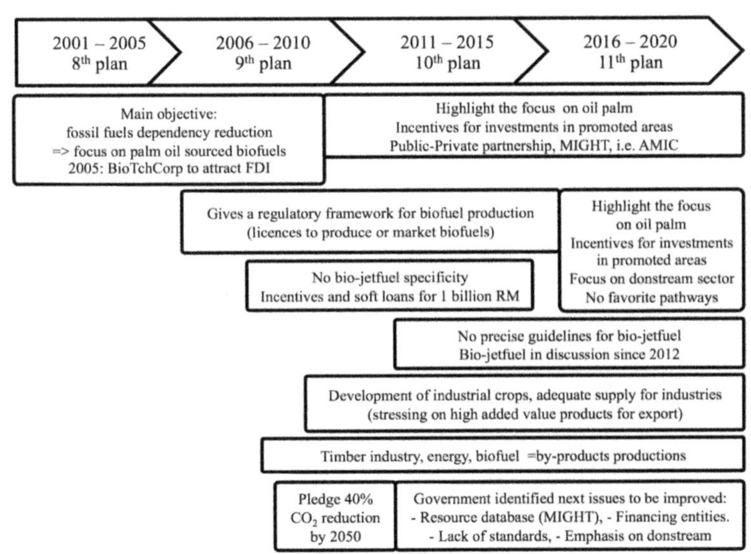

Figure 36: Outcomes linked to biomass and bio-jetfuel.

Conclusion and perspectives

The civil aviation industry targets 50% reduction in net CO_2 emission by 2050, and will need at least 2 million tonnes of biofuel by 2020. To date there are only three certified processes that can produce Jet A1 kerosene from biomass. One is the HEFA process, mainly using fatty biomass. The two others that use sugars, starch and lignocellulosic materials (polymers of sugars) are the biomass-to-liquid, via the Fischer-Tropsch process, and the Synthesised Iso-Paraffinic fuel, via fermentation. In South East Asia, biomass grows fast and in abundance, both in agro-systems and in natural ecosystems such as the rainforest.

There lies the possible barrier for jetfuel from biomass. It can encourage deforestation if unsustainably feeding off directly (timber) or indirectly (dedicated energy crops that need space, thus encouraging deforestation) on the forests. It can also increase the possibility of food shortage, especially among the poorest populations by encouraging the development of energy crops in competition with food crops.

South East Asia is where the aviation demand and the projected demand for biomass energy will grow the fastest. Three major international air-hubs are located in or at the doorstep of Malaysia: Bangkok in Thailand, Kuala Lumpur in Peninsular Malaysia, and Singapore across a bridge at the southern tip of the peninsula. But South East Asia is also one of the regions in the world that has less potential for expansion of new crop areas. This creates a "scissors effect", leaving little options except to depend on lignocellulosic materials, the basic component of all plants, and one of the most abundant biopolymers in the world. It constitutes the bulk of all forest products and agricultural wastes.

In Malaysia the major source of vegetable oil is oil palm, but the international prices of this commodity and its fatty residues, make it economically out of reach of major biofuel uses, if not subsided. Despite the existence of environmental certification for this crop, it is not completely safe from possible environmental criticisms, due to the connection of its value chain with the rainforests that surround the plantations. The major sources of lignocellulosic residues are oil palm, forestry, paddy, coconut, sugarcane, and rubber plantations. Despite the dominance of oil palm and forest areas, and the relative productivity of paddy, no crop by itself in Peninsular Malaysia would be sufficient to constitute a sustainable source of residues for bio-jetfuel, at the scale potentially demanded by one of the three hubs close to this region. The total potential of a multi-crop sourcing strategy could provide up to 50 million tonnes of lignocellulosic fresh residues from Peninsular Malaysia, and 100 million tonnes from the whole of Malaysia. Similarly a conservative multi-crop strategy, that avoids the oil palm sector, could provide up to 7-8 million tonnes of lignocellulosic fresh residues from Peninsular Malaysia, and 15 million tonnes from the whole Malaysia. These numbers are indicative and subject to variance of more than plus or minus 50%.

The specific configuration of Malaysia, in two land masses separated by the South China Sea, and the scattered nature of all the crops, in a mosaic of land uses spread across a contrasted geography, creates a challenge for the transport cost of the raw materials. Initial simulations on one crop shows the necessity of integrating the collection of raw material and of selecting carefully the place(s) for biorefinery(ies). Adding the sources of Sabah and Sarawak to those from the peninsula is not impossible, but should be optimised. It implies a multi-mode transport system with one segment by sea - and according to the harbour facilities or the volumes and scales, the economic models need to be assessed and weighted, and if found wanting, rejected. Simultaneously adding the volumes available on a regional scale will drastically change the scale of the operations and may optimise the cost.

A series of incentives and institutional tools exist to support biofuels and biotechnology in a few directions, but in a non-systematic way, with little consolidation for effective applications or large-scale implementation. A clear orientation from the public policy is to push for oil palm sector related investments, but it might not be an advantage for specific companies, which may be afraid to be linked with the global environmental polemics on oil palm environmental effects. Other policies promote other sectors as well, leaving largely enough room for any other biomass strategy to take off, provided governmental policies and offices consolidate their actions.

One new institutional tool is the Aerospace Malaysia Innovation Centre, with its Center of Excellence on Biomass Valorisation for Aviation, whose job is to assess local solutions for sustainable biomass production. The aim is to determine the most suitable feedstocks to ensure future jet fuel feedstocks is based only on sustainable solutions. Their next step, after this initial statement on the potentialities, will be to build and assess production scenarios, for the policy makers to select among a portfolio of sustainable options for a realistic and sustainable roadmap.

An incremental approach would ideally consider a careful assessment of optimal feedstock to chose in accordance to their location, scale of supply, and other technical, environmental and economic parameters. For example if one needs to get biomass from Sabah and Sarawak in order to supplement the biomass available in the peninsula, it can be as profitable if not more, to also source from Sumatra, Sulawesi, etc. These factors call for a regional approach to the supply pools. Regional economics and regional scale of analysis are also often the best scales in order to ensure sound sustainability assessment.

At such scales, scenarios to build synergies between international hubs, or to measure their potential competition and cannibalism, can be envisioned. Ideal or alternative location of bio-refineries can be computed; several industrial strategies and scenarios can be compared, along with appropriate economic models. There can be several different economic models according to the end-users. Different airlines could implement bio-jetful according to different time scales. Their customer base can be different with different cultural backgrounds, thus they could favour different kinds of ideal biomass feedstock. Companies running the biorefineries have their own economic models too, with different long-term strategies. These factors, too, are critical when weighting alternative scenarios or mill locations. For the Malaysian government or its neighbours, a set of clear institutional tools and criteria for promoting or managing the dynamic of bio-jetfuls in the region should be presented, and assessed each with their advantages and disadvantages. All of these elements are necessary components for the institutions and the companies to build and negotiate the roadmap(s) that will lead to sustainable development of bio-jetful in South East Asia.

References

References specifically relevant to Malaysia

Ab Rahman, A., 2013. Soil resource management and conservation division. Presented at the Department of Agriculture Meetings, Kuala Lumpur, Malaysia, pp. 1-30.

Abu Bakar, N., 2013. Country Presentation on Status of Bioenergy Development In Malaysia. Presented at the International Energy Agency Conference, pp. 1-14.

Anonymous, 2015. Malaysia Jet fuel consumption [WWW Document]. theglobaleconomy.com. URL http://www.theglobaleconomy.com/Malaysia/jet_fuel_consumption/ (accessed 8.13.15).

Anonymous, 2014a. The Nexus of Biofuels, Climate Change, and Human Health: Workshop Summary. National Academies Press, Washington DC, USA.

Anonymous, 2014b. Airbus joins biofuel effort in Malaysia. Air Cargo Week 17, 1.

Anonymous, 2014c. Palm Oil facts and figures, simedarby.com. Kuala Lumpur, Malaysia.

Anonymous, 2013a. National Biomass Strategy 2020: New wealth creation for Malaysia's biomass industry. Agensi Inovasi Malaysia, Kuala Lumpur, Malaysia.

Anonymous, 2013b. Southeast Asia energy outlook. IEA Publications, Paris, France.

Anonymous, 2013c. A feasibility study of Australian feedstock and production capacity to produce sustainable aviation fuel. Qantas Airways Ltd, Mascot, Australia.

Anonymous, 2013d. Case Study: The Palm Oil Example, in: Pool, R. (Ed.), The Nexus of Biofuels, Climate Change, and Human Health: Workshop Summary. Washington DC, USA, pp. 29-36.

Anonymous, 2012. IATA 2012 Report on Alternative Fuels. IATA, Montreal, Canada.

Anonymous, 2011a. Renewable Fuel Standard: Potential Economic and Environmental Effects of U.S. Biofuel Policy. National Academies Press, Washington DC, USA.

Anonymous, 2011b. Laporan banci pertubuhan pertanian - Report on the census of agricultural establishments, dropbox.com. Putrajaya, Malaysia.

Anonymous, 2011c. Flight Path to Sustainable Aviation: Towards Establishing a Sustainable Aviation Fuels Industry in Australia and New Zealand. CSIRO.

Anonymous, 2011d. Biofuels Vital Graphics: Powering a green economy. UNEP, Paris, France.

Anonymous, 2011e. Delivering the Future: Global Market Forecast 2011-2013. Airbus group, Blagnac, France.

Anonymous, 2007a. Peninsular Malaysia - federal ports. Department of Statistics, Malaysia.

Anonymous, 2007b. Peninsular Malaysia - International airports, domestic airports and airstrips. Department of Statistics, Malaysia.

Anonymous, 2007c. Bioenergy project development & biomass supply. International Energy Agency, Paris, France.

Anonymous, 2007d. A Guide to Airports in Asia Pacific. KPMG International.

Balan, V., 2014. Current Challenges in Commercially Producing Biofuels from Lignocellulosic Biomass 2014, 1-31. doi:10.1155/2014/463074.

bin Wan Abdullah, W.A.A., 2014. Malaysia airports annual report 2013: Accelerating growth momentum. Malaysia Airports Holdings Berhad, Sepang, Malaysia.

binti Jamil, N.H., 2009. Biomass potential energy from agricultural wastes. University Malaysia Sarawak, Kuching.

binti Yusof, A., 2011. Analysis of renewable energy potential in Malaysia / Adawati binti Yusof. Universiti Malaya, Kuala Lumpur, Malaysia.

Brown, B., 2011. Economic Benefits from Air Transport in Malaysia. Oxford Economics, Oxford.

Charles, M.B., Barnes, P., Ryan, N., Clayton, J., 2007. Airport futures: Towards a critique of the aerotropolis model. Futures 39, 1009-1028. doi:10.1016/j.futures.2007.03.017.

Daniel, L., 2012. Valorisation of Indonesian Plant Oil Resources. Bandung, Indonesia.

de Andrade, R.M.T., Miccolis, A., 2011. Policies and institutional and legal frameworks in the expansion of Brazilian biofuels. Cifor, Bogor, Indonesia.

Ensor, J.D., 2004. Malaysia Transport Pricing Strategies, Measures, and Policies: Inception Report, Cambridge. Cambridge.

Erb, K.-H., Krausmann, F., Lucht, W., Haberl, H., 2009. Embodied HANPP: Mapping the spatial disconnect between global biomass production and consumption. Ecological Economics 69, 328-334. doi:10.1016/j.ecolecon.2009.06.025.

Fritsche, U.R., Hünecke, K., Hermann, A., Schulze, F., Wiegmann, K., 2006. Sustainability standards for bioenergy. Berlin, Germany.

Goh, C.S., Lee, K.T., 2011. Second-generation biofuel (SGB) in Southeast Asia via lignocellulosic biorefinery: Penny-foolish but pound-wise. Renewable and Sustainable Energy Reviews 15, 2714-2718. doi:10.1016/j.rser.2011.02.036.

Goh, C.S., Lee, K.T., 2010. Palm-based biofuel refinery (PBR) to substitute petroleum refinery: An energy and emergy assessment. Renewable and Sustainable Energy Reviews 14, 2986-2995. doi:10.1016/j.rser.2010.07.048.

Griffin, W.M., Michalek, J., Matthews, H.S., Hassan, M., 2014. Availability of biomass residues for co-firing in Peninsular Malaysia: Implications for cost and GHG emissions in the electricity sector. Energies 7, 804-823. doi:10.3390/en7020804.

Gumartini, T., 2009. Biomass energy in the Asia-Pacific region: current status, trends and future setting. Asia-Pacific Forestry Sector Outlook Study II.

Hong, T.D., Soerawidjaja, T.H., Reksowardojo, I.K., Fujita, O., Duniani, Z., Pham, M.X., 2013. A study on developing aviation biofuel for the Tropics: Production process—Experimental and theoretical evaluation of their blends with fossil kerosene. Chemical Engineering and Processing Process Intensification 74, 124-130. doi:10.1016/j.cep.2013.09.013.

Huber, G.W., Iborra, S., Corma, A., 2006. Synthesis of Transportation Fuels from Biomass: Chemistry, Catalysts, and Engineering. Chemical Reviews 106, 4044-4098. doi:10.1021/cr068360d.

Klass, D.L., 2004. Biomass for Renewable Energy and Fuels. Encyclopedia of Energy 1, 193-212. doi:10.1016/b0-12-176480-x/00353-3.

Kline, K.L., Oladosu, G.A., Wolfe, A.K., Perlack, R.D., Dale, V.H., McMahon, M., 2008. Biofuel feedstock assessment for selected countries (No. ORNL/TM-2007/224). Springfield, USA.

Lee, K.T., Ofori-Boateng, C., 2013. Sustainability of Biofuel Production from Oil Palm Biomass. Springer Science & Business Media. doi:10.1007/978-981-4451-70-3.

Li, Z., Fox, J.M., 2012. Mapping rubber tree growth in mainland Southeast Asia using time-series MODIS 250 m NDVI and statistical data. Applied Geography 32, 420-432. doi:10.1016/j.apgeog.2011.06.018.

Lim, S., Teong, L.K., 2010. Recent trends, opportunities and challenges of biodiesel in Malaysia: An overview. Renewable and Sustainable Energy Reviews 14, 938-954. doi:10.1016/j.rser.2009.10.027.

Luque, R., Herrero-Davila, L., Campelo, J.M., Clark, J.H., Hidalgo, J.M., Luna, D., Marinas, J.M., Romero, A.A., 2008. Biofuels: a technological perspective. Energy & Environmental Science 1, 542. doi:10.1039/b807094f.

Malins, C., Searle, S., Baral, A., 2014. A Guide for the Perplexed to the Indirect Effects of Biofuels Production, ICCT.

Nakada, S., Saygin, D., Gielen, D., 2014. Global Bioenergy Supply and Demand Projections. International Renewable Energy Agency, Abu Dhabi, UAE.

Ong, H.C., Mahlia, T., Masjuki, H.H., 2011. A review on energy scenario and sustainable energy in Malaysia. Renewable and Sustainable Energy Reviews 15, 639-647. doi:10.1016/j.rser.2010.09.043.

Pang, T.W., Bilson, K., Lee, M.T., 2013. Biorefinery potentials in Sabah. Presented at the Bio Borneo Conference, Kota Kinabalu, Malaysia, pp. 1-39.

Pang, T.W., Lee, M.T., 2014. The role of POIC Lahad Datu in industrializing Sabah and how geospatial and surveying communities can contribute. Presented at the Sabah international surveyors congress, Koata Kinabalu, Malaysia, pp. 1-29.

Pool, R., 2013. The Nexus of Biofuels, Climate Change, and Human Health: Workshop Summary. National Academies Press, Washington DC, USA.

Rajgor, G., 2011. Biofuels bottleneck. Renewable Energy Focus 12, 66-71. doi:10.1016/s1755-0084(11)70158-2.

Ravindranath, N.H., Lakshmi, C.S., Manuvie, R., 2011. Biofuel production and implications for land use, food production and environment in India. Energy Policy 39, 5737-5745. doi: 10.1016/j.enpol.2010.07.044.

Sayer, J., Ghazoul, J., Nelson, P., Boedhihartono, A.K., 2012. Oil palm expansion transforms tropical landscapes and livelihoods. Global Food Security 1, 114-119. doi:10.1016/j.gfs.2012.10.003

Sgouridis, S.P., 2003. Freight transportation in Malaysia: Technological and organizational issues from an ITS perspective (No. AY 2002/2003 Spring Inception Report). Massachusetts Institute of Technology.

Shafie, S.M., Mahlia, T., Masjuki, H.H., 2012. A review on electricity generation based on biomass residue in Malaysia. Renewable and Sustainable Energy Reviews 16, 5879-5889. doi: 10.1016/j.rser.2012.06.031.

Stichnothe, H., Schuchardt, F., 2011. Life cycle assessment of two palm oil production systems. Biomass and Bioenergy 35, 3976-3984. doi:10.1016/j.biombioe.2011.06.001.

Syed, A., Nowakowski, D., Nicholson, P., Ahmad, S., 2014. Australian liquid fuels technology assessment 2014. Bureau of Resources and Energy Economics, Canberra, Australia.

Tahir, T.A., 2012. Waste audit at coconut-based industry and vermicomposting of different types of coconut waste. University of Malaya, Kuala Lumpur, Malaysia.

Timilsina, G.R., Shrestha, A., 2011. How much hope should we have for biofuels? Energy. doi: 10.1016/j.energy.2010.08.023.

Vakkilainen, E., Kuparinen, K., Heinimo, J., 2013. Large industrial users of energy biomass. IEA Bioenergy, Lappeenranta, Finland.

van Gelder, J.W., German, L., 2011. Biofuel finance: global trends in biofuel finance in forest-rich countries of Asia, Africa and Latin America and implications for governance. CIFOR infobriefs 36, 1-12.

Yevich, R., Logan, J.A., 2003. An assessment of biofuel use and burning of agricultural waste in the developing world. Global Biogeochem. Cycles 17, 6.1-6.21. doi:10.1029/2002gb001952.

General references

Abadi, A., Farquharson, B., 2012. Integrated farming systems in a changing climate. Future Farm 2012, 8-9.

Alexander, D., Alexander, K., 2012. Mallees prove double-edged sword. Future Farm 2012, 4-5.

Alfstad, T., 2008. World Biofuels Study: Scenario Analysis of Global Biofuel Markets. Brookhaven National Laboratory, Upton, USA.

Altman, R., 2012. Sustainable aviation fuel, supporting, complementing, maritime fuel developments, Presented at the Sustainable Maritime Fuels Forum, Sydney, Australia, pp. 1-24.

Anonymous, 2014a. Mallee Jet Fuel at Perth Airport 2021.

Anonymous, 2014b. Biomass meeting the biotechnology challenge. Total, Paris, France.

Anonymous, 2014c. Lazard's levelized cost of energy analysis. Hamilton, USA.

Anonymous, 2013a. Flightpath to aviation BioFuels in Brazil: action plan. Sao Paolo, Brazil.

Anonymous, 2013b. Biofuels and Food Security Issues, in: Pool, R. (Ed.), The Nexus of Biofuels, Climate Change, and Human Health: Workshop Summary. pp. 121-136.

Anonymous, 2013c. NREL Proves Cellulosic Ethanol Can Be Cost Competitive (Fact Sheet) (No. NREL/FS-6A42-60663). National Renewable Energy Laboratory, Golden, USA.

Anonymous, 2012a. Biomass for Power Generation, Renewable energy technologies cost analysis series. Bonn, Germany.

Anonymous, 2012b. Climate Change and Aviation: Issues, Challenges and Solutions. Routledge.

Anonymous, 2012c. Mallee Biomass for biofuels.

Anonymous, 2011. ICAO environmental report 2010. ICAO.

Anonymous, 2010a. Alternative Fuels, in: Transitions to Alternative Vehicles and Fuels. National Academies Press, Washington DC, USA, pp. 42-76.

Anonymous, 2010b. Fuels, in: Transitions to Alternative Vehicles and Fuels. National Academies Press, Washington DC, USA, pp. 305-330.

Anonymous, 2010c. Modeling, in: Transitions to Alternative Vehicles and Fuels. National Academies Press, Washington DC, USA, pp. 331-378.

Anonymous, 2010d. Modeling the Transition to Alternative Vehicles and Fuels, in: Transitions to Alternative Vehicles and Fuels. National Academies Press, Washington DC, USA, pp. 89-130.

Anonymous, 2010e. Policy Options, in: Transitions to Alternative Vehicles and Fuels. National Academies Press, Washington DC, USA, pp. 152-160.

Anonymous, 2010f. Transitions to Alternative Vehicles and Fuels. National Academies Press, Washington DC, USA.

Anonymous, 2009a. Converting Waste Agricultural Biomass into a Resource: Compendium of Technologies (No. DTI/1203/JP). Osaka, Japan.

Anonymous, 2009b. Liquid Transportation Fuels from Coal and Biomass: Technological Status, Costs, and Environmental Impacts. Washington DC, USA.

Anonymous, 2008a. Bioethanol production, in: Bioethanol Sugarcane Based Energy for Sunstainable Development. BNDES, Sao Paolo, Brazil, pp. 63-97.

Anonymous, 2008b. Water Implications of Biofuels Production in the United States. Washington DC, USA.

Anonymous, 2008c. The state of food and agriculture - Biofuels: prospects, risks and opportunitie. FAO, Rome, Italy.

Anonymous, 2007. Aviation fuels: technical review (No. FTR-3). Houston, USA.

Anonymous, 2000. Biobased Industrial Products: Research and Commercialization Priorities. National Academies Press, Washington DC, USA.

Anonymous, 1999. Review of the Research Strategy for Biomass-Derived Transportation Fuels. National Academies Press, Washington DC, USA.

Anonymous, 1990. Fuels to Drive Our Future, Energy Sources. National Academies Press, Washington DC, USA.

Anwar, Z., Gulfraz, M., Irshad, M., 2014. Agro-industrial lignocellulosic biomass a key to unlock the future bio-energy: A brief review. Journal of Radiation Research and Applied Sciences 7, 163-173. doi:10.1016/j.jrras.2014.02.003.

Ausubel, J.H., Herman, R., 1988. Cities and Their Vital Systems: Infrastructure Past, Present, and Future. National Academies Press, Washington DC, USA.

Bai, Y., Hwang, T., Kang, S., Ouyang, Y., 2011. Biofuel refinery location and supply chain planning under traffic congestion. Transportation Research Part B: Methodological 45, 162-175. doi:10.1016/j.trb.2010.04.006.

Bloyd, C., Foster, N., 2014. An Update on Ethanol Production and Utilization in Thailand—2014 (No. PNNL-23673). Pacific Nortwest National Laboratory, Springfield, USA. doi:10.1787/agr_outlook-2014-graph71-en.

Börjesson, P., Gustavsson, L., 1996. Regional production and utilization of biomass in Sweden. Energy 1996, 1033-1036. doi:10.1016/0960-1481(96)88456-4.

Buijs, N.A., Siewers, V., Nielsen, J., 2013. Advanced biofuel production by the yeast Saccharomyces cerevisiae. Curr Opin Chem Biol 17, 480-488. doi:10.1016/j.cbpa.2013.03.036.

Cafferty, K.G., Muth, D.J., Jacobson, J.J., Bryden, K.M., 2013. Model based biomass system design of feedstock supply systems for bioenergy production (No. DETC2013-13559). ASME, Portland.

Carr, B., 2013. Breaking the Barriers with Breakthrough Jet Fuel Solutions, Presented at the Renewable Aviation Fuel Joint Development Program Paris Air Show, Le Bourget, pp. 1-15.

Caslin, B., Finnan, J., 2010. Straw for Energy. Tillage.

Chen, J.Y., Ma, X.J., Feng, X.Y., Han, X.L., 2012. Status study of aviation biofuel at home and abroad. Renewable Energy.

Cherry, J., 2010. Turning the promise of synthetic biology into commercial reality for health and energy, Presented at the European Commission Synthetic Biology Workshop, Brussels, Belgium, pp. 1-21.

Chiaramonti, D., Prussi, M., Buffi, M., Tacconi, D., 2014. Sustainable bio kerosene: Process routes and industrial demonstration activities in aviation biofuels. Applied Energy 136, 767-774. doi:10.1016/j.apenergy.2014.08.065.

Chuck, C.J., Donnelly, J., 2014. The compatibility of potential bioderived fuels with Jet A-1 aviation kerosene. Applied Energy 118, 83-91. doi:10.1016/j.apenergy.2013.12.019.

Clarke, S., Eng, P., Preto, F., 2011. Biomass Densification for Energy Production.

Collett, J.R., Meyer, P.A., Jones, S.B., 2014. Preliminary Economics for Hydrocarbon Fuel Production from Cellulosic Sugars (No. PNNL-23374). Pacific Northwest National Laboratory, Springfield, USA.

Connelly, E.B., Colosi, L.M., Clarens, A.F., Lambert, J.H., 2015. Risk Analysis of Biofuels Industry for Aviation with Scenario-Based Expert Elicitation. Systems Engineering 18, 178-191. doi: 10.1002/sys.21298.

Cremonez, P.A., Feroldi, M., de Araújo, A.V., Borges, M.N., Meier, T.W., Feiden, A., Teleken, J.G., 2015. Biofuels in Brazilian aviation: Current scenario and prospects. Renewable and Sustainable Energy Reviews 43, 1063-1072. doi:10.1016/j.rser.2014.11.097.

Cremonez, P.A., Feroldi, M., de Jesus de Oliveira, C., Teleken, J.G., Alves, H.J., Sampaio, S.C., 2013. Environmental, economic and social impact of aviation biofuel production in Brazil. N Biotechnol -. doi:10.1016/j.nbt.2015.01.001.

Davis, R., Biddy, M., Tan, E., Tao, L., Jones, S., 2013a. Biological Conversion of Sugars to Hydrocarbons Technology Pathway (No. NREL/TP-5100-58054). National Renewable Energy Laboratory, Springfield, USA.

Davis, R., Tao, L., Tan, E.C.D., Biddy, M.J., Beckham, G.T., Scarlata, C., Jacobson, J., Cafferty, K., Ross, J., Lukas, J., Knorr, D., Schoen, P., 2013b. Process Design and Economics for the Conversion of Lignocellulosic Biomass to Hydrocarbons: Dilute-Acid and Enzymatic Deconstruction of Biomass to Sugars and Biological Conversion of Sugars to Hydrocarbons (No. NREL/TP-5100-60223). National Renewable Energy Laboratory, Springfield, USA.

Davis, S.C., 1999. Transportation Energy Data Book, Edition 19 (No. ORNL-6958). Oak Ridge National Laboratory, Oak Ridge, USA.

de Jong, B., Siewers, V., Nielsen, J., 2012. Systems biology of yeast: enabling technology for development of cell factories for production of advanced biofuels. Current Opinion in Biotechnology 23, 624-630. doi:10.1016/j.copbio.2011.11.021.

De Wit, M., Faaij, A., 2010. European biomass resource potential and costs. Biomass and Bioenergy 34, 188-202. doi:10.1016/j.biombioe.2009.07.011.

Demirbas, A., 2011. Competitive liquid biofuels from biomass. Applied Energy 88, 17-28. doi: 10.1016/j.apenergy.2010.07.016.

Demirbas, A., 2008. Biofuels sources, biofuel policy, biofuel economy and global biofuel projections. Energy Conversion and Management 49, 2106-2116. doi:10.1016/j.enconman. 2008.02.020

Demirbas, A., 2007. Progress and recent trends in biofuels. Progress in Energy and Combustion Science 33, 1-18. doi:10.1016/j.pecs.2006.06.001.

Diederichs, W., Gorgens, J.F., 2015. Techno-Economic Assessment of Processes that produce Jet Fuel from Plant-Derived Sources, Presented at the Renewable energy postgraduate symposium, Stellenbosch, South Africa, pp. 1-15.

Doat, J., 1977. Le pouvoir calorifique des bois tropicaux. Revve Bois et Forests des Tropiques 172, 33-55.

Dubbelboer, M., 2009. Bio fuels: A review and comparison of available bio fuels, techniques and their environmental, economical, political and social consequences. Afstudeeropdracht Faculteit Natuurwetenschappen, Groningen, Netherlands.

Dunnett, A.J., Adjiman, C.S., Shah, N., 2008. A spatially explicit whole-system model of the lignocellulosic bioethanol supply chain: an assessment of decentralised processing potential. Biotechnology for Biofuels 1, 13. doi:10.1186/1754-6834-1-13

Elgowainy, A., Han, J., Wang, M., Carter, N., Stratton, R., Hileman, J., Malwitz, A., Balasubramanian, S., 2012. Life-cycle analysis of alternative aviation fuels in GREET (No. ANL/ESD/12-8). Argonne National Laboratory, Oak Ridge, USA.

Eychenne, F., 2013. AIRBUS and Alternative Fuels: A practical Approach, Presented at the Airbus Environmental Affairs, Toulouse, France, pp. 1-10.

Flintsch, G.W., 2008. Life-Cycle Assessment as a tool for sustainable transportation infrstructure management, in: Kutz, M. (Ed.), Environmentally Conscious Transportation. John Wiley & Sons, pp. 257-282.

Gegg, P., Budd, L., Ison, S., 2014. The market development of aviation biofuel: Drivers and constraints. Journal of Air Transport Management 39, 34-40. doi:10.1016/j.jairtraman.2014.03.003

Gnansounou, E., Dauriat, A., 2010. Techno-economic analysis of lignocellulosic ethanol: A review. Bioresource Technology 101, 4980-4991. doi:10.1016/j.biortech.2010.02.009.

Gomez, L.D., Steele King, C.G., McQueen Mason, S.J., 2008. Sustainable liquid biofuels from biomass: the writing's on the walls. New Phytologist 178, 473-485. doi:10.1111/j.1469-8137.2008.02422.x.

Goss, K., 2011. Mallee biomass takes to the skies. Australian Forest Grower 34, 32.

Goss, K., Abadi, A., Crossin, E., Stuckley, C., Turnbull, P., 2014. Sustainable Mallee Jet Fuel. AIRBUS; Future Farm, Crawley, Australia.

Graham, P., Reedman, L., Philips, K., Morozow, O., Wallace, V., Poldy, F., Maher, M., Davis, L., 2008a. Fuel for Thought - the future of transport fuels: challenges and opportunities. Commonwealth Scientific and Industrial Research Organisation, Campbell, Australia.

Graham, P., Reedman, L., Poldy, F., 2008b. Modelling of the Future of Transport Fuels in Australia (No. IR 1046). CSIRO, Newcastle. Australia.

Graham, P., Reedman, L., Rodriguez, L., Raison, J., Braid, A., Haritos, V., Brinsmead, T., Hayward, J., Taylor, J., OConnell, D., Adams, P., 2011. Sustainable aviation fuels road map: data assumptions and modelling. CSIRO, Newcastle, Australia.

Granda, C.B., Zhu, L., Holtzapple, M.T., 2007. Sustainable liquid biofuels and their environmental impact. Environmental Progress 26, 233-250. doi:10.1002/ep.10209.

Gray, D., Sato, S., Garcia, F., Eppler, R., Cherry, J., 2014. Amyris, Inc. Integrated Biorefinery Project Summary (No. EE0002869). Amyris.

Gresshoff, P., 2014. The contrasting need for food and biofuel: Can we afford biofuel?, in: Sykes, H. (Ed.), A Love of Ideas. Future Leaders, Sydney, Australia, pp. 114-152.

Grosshans, R., Kostelnik, K.M., Jacobson, J.J., 2007. Sustainable Harvest for Food and Fuel (No. INL/CON-07-12655). Idaho Falls, USA.

Gu, T., 2013. Green Biomass Pretreatment for Biofuels Production. Springer Science & Business Media, Athens, USA.

Guariguata, M.R., Masera, O.R., Johnson, F.X., Maltitz, von, G., Bird, N., Tella, P., Martinez-Bravo, R., 2011. A review of environmental issues in the context of biofuel sustainability frameworks. CIFOR Occasional Papers, Bogor.

Hamelinck, C.N., Faaij, A.P.C., 2006. Outlook for advanced biofuels. Energy Policy 34, 3268-3283. doi:10.1016/j.enpol.2005.06.012.

Haq, Z., 2013. Status of Biofuels Activities at DOE, Presented at the Society of Automotive Engineers SAE GovernmentIndustry Meeting, pp. 1-21.

Hileman, J., 2013. Overview of FAA Alternative Jet Fuel Activities, Presented at the Biomass R&D Technical Advisory Committee, Washington D.C., pp. 1-20.

Hnain, A.K., Cockburn, L.M., Lefebvre, D.D., 2011. Microbiological processes for waste conversion to bioenergy products: Approaches and directions. Environ. Rev. 19, 214-237. doi:10.1139/a11-007

Hoefnagels, R., Smeets, E., Faaij, A., 2010. Greenhouse gas footprints of different biofuel production systems. Renewable and Sustainable Energy Reviews 14, 1661-1694.

Holladay, J., Albrecht, K., Hallen, R., 2014. Renewable routes to jet fuel, Presented at the Japan Aviation Environmental WorkshopInnovative concepts for carbon neutral growth, pp. 1-44.

Hu, X., Qi, P., Fu, X., He, H., HUANG, G., 2012. Technology development background and application status on aviation biofuels. Chemical Industry and Engineering Progress.

Huang, Y., Chen, C.-W., Fan, Y., 2010. Multistage optimization of the supply chains of biofuels. Transportation Research Part E: Logistics and Transportation Review 46, 820-830. doi: 10.1016/j.tre.2010.03.002.

Izaurralde, R.C., Sahajpal, R., Zhang, X., Manowitz, D.H., 2012a. National Geo-Database for Biofuel Simulations and Regional Analysis (No. PNNL-21284). Pacific Northwest National Laboratory, Richland, USA.

Izaurralde, R.C., Sahajpal, R., Zhang, X., Manowitz, D.H., 2012b. National Geo-Database for Biofuel Simulations and Regional Analysis of Biorefinery Siting Based on Cellulosic Feedstock Grown on Marginal Lands (No. PNNL-21283). Pacific Northwest National Laboratory, Richland, USA.

Jacobson, J.J., Jeffers, R.F., 2013. Bioenergy market competition for biomass: a system dynamics review or current policies (No. INL/CON-13-28535). Idaho National Laboratory, Idaho Falls, USA.

Jacobson, J.J., Roni, M., Lamers, P., Cafferty, K.G., 2014. Biomass Feedstock Supply System Design and Analysis (No. INL/EXT-14-32377). Idaho National Laboratory, Idaho Falls, USA.

Jacobson, J.J., Searcy, E., Muth, D., Wilkerson, E., Sokansanj, S., Jenkins, B., Tittman, P., Parker, N., Hart, Q., Nelson, R., 2009. Sustainable biomass supply systems (No. INL/CON-09-15568), Idaho National Laboratory, Idaho Fals, USA.

Johnson, C., Hettinger, D., 2014. Geography of Existing and Potential Alternative Fuel Markets in the United States (No. NREL/TP-5400-60891). National Renewable Energy Laboratory, Golden, USA.

Johnson, M., Schenk, D., Miller, B., Altman, R., Brand, M., McDonald, A., Thompson, T., Driver, J., Leistritz, L., Leholm, A., Hodur, N., David, P., Glassman, D., Anumakonda, A., 2012. Guidelines for Integrating Alternative Jet Fuel Into the Airport Setting. Transportation Research Board, Washington DC, USA.

Jovanovic Tews, I., Jones, S.B., Santosa, D.M., Dai, Z., Ramasamy, K.K., Zhu, Y., 2010. A survey of Opportunities for Microbial Conversion of Biomass to Hydrocarbon Compatible Fuels. doi: 10.2172/1013947.

Kalinauskaitė, S., Sakalauskas, A., 2013. Relation of energy content variations of straw to the fraction size, humidity, composition and environmental impact. Agronomy Research.

Karatzos, S., McMillan, J.D., Saddler, J.N., 2014. The Potential and Challenges of Drop-in Biofuels (No. T39-T1). IEA Bioenergy.

Kenney, K.L., Hess, J.R., Smith, W.A., Muth, D.J., 2012. Improving Biomass Logistics Cost within Agronomic Sustainability Constratints and Biomass Quality Targets (No. INL/CON-12-25427). Idaho National Laboratory, Idaho Falls, USA.

King, D., Inderwildi, O.R., Williams, A., Hagan, A., Löffler, K., Gillman, N., Weihe, U., Oertel, S., 2010. The Future of Industrial Biorefineries (No. 210610). World Economic Forum, Geneva, Switzerland.

Kivits, R., Charles, M.B., Ryan, N., 2010. A post-carbon aviation future: Airports and the transition to a cleaner aviation sector. Futures. doi:10.1016/j.futures.2009.11.005.

Korpinen, O.-J., Jäppinen, E., Ranta, T., 2013. A Geographical-Origin-Destination Model for Calculating the Cost of Multimodal Forest-Fuel Transportation. JGIS 05, 96-108. doi: 10.4236/jgis.2013.51010.

Koshel, P., McAllister, K., 2010. Expanding Biofuel Production: Sustainability and the Transition to Advanced Biofuels: Summary of a Workshop. National Academies Press, Washington DC, USA.

Krammer, P., Dray, L., Köhler, M.O., 2013. Climate-neutrality versus carbon-neutrality for aviation biofuel policy. Transportation Research Part D: Transport and Environment 23, 64-72. doi: 10.1016/j.trd.2013.03.013.

Krausmann, F., Erb, K.-H., Gingrich, S., Lauk, C., Haberl, H., 2008. Global patterns of socioeconomic biomass flows in the year 2000: A comprehensive assessment of supply, consumption and constraints. Ecological Economics 65, 471-487. doi:10.1016/j.ecolecon. 2007.07.012.

Lam, H.L., Varbanov, P., Klemeš, J., 2010. Minimising carbon footprint of regional biomass supply chains. Resources, Conservation and Recycling 54, 303-309. doi:10.1016/j.resconrec. 2009.03.009.

Larasati, A., Liu, T., Epplin, F.M., 2012. An Analysis of Logistic Costs to Determine Optimal Size of a Biofuel Refinery. Engineering Management Journal. doi: 10.1080/10429247.2012.11431956

Lave, L.B., Griffin, W.M., 2008. THE ECONOMIC AND ENVIRONMENTAL FOOTPRINTS OF TRANSPORTATION, in: Kutz, M. (Ed.), Environmentally Conscious Transportation. John Wiley & Sons, pp. 1-13.

Leary, N., 1999. Aviation and the Global Atmosphere: A Special Report of IPCC Working Groups I and III, Presented at the IPCC Symposium, Tokyo, Japan, pp. 1-7.

Lee, J., Mo, J., 2011. Analysis of Technological Innovation and Environmental Performance Improvement in Aviation Sector. International Journal of Environmental Research and Public Health 8, 3777-3795. doi:10.3390/ijerph8093777.

Lee, S.K., Chou, H., Ham, T.S., Lee, T.S., 2008. Metabolic engineering of microorganisms for biofuels production: from bugs to synthetic biology to fuels. Current Opinion in Biotechnology 19, 556-563. doi:10.1016/j.copbio.2008.10.014.

Li, Q., Song, J., Peng, S., Wang, J.P., Qu, G.-Z., Sederoff, R.R., Chiang, V.L., 2014. Plant biotechnology for lignocellulosic biofuel production. Plant Biotechnology Journal 12, 1174-1192. doi:10.1111/pbi.12273.

Mabee, W.E., McFarlane, P.N., Saddler, J.N., 2011. Biomass availability for lignocellulosic ethanol production. Biomass and Bioenergy 35, 4519-4529. doi:10.1016/j.biombioe.2011.06.026

Malina, R., 2014. Innovative Technologies for Alternative Aviation Fuels, Presented at the Sustainable Aviation Fuels Forum, Madrid, Spain, pp. 1-14.

Malina, R., Pearlson, M., Carter, N., Bredehoeft, M., Wollersheim, C., Olcay, H., Hileman, J., Barrett, S., 2012. HEFA and F-T jet fuel cost analyses, Presented at the Laboratory for Aviation and the Environment Presentations, pp. 1-35.

Manicore, J.-M., 2015. Could we totally substitute oil by biofuels? [WWW Document]. Manicore. URL http://www.manicore.com (accessed 2.27.15).

Martinkus, N., Shi, W., Lovrich, N., Pierce, J., Smith, P., Wolcott, M., 2014. Integrating biogeophysical and social assets into biomass-to-biofuel supply chain siting decisions. Biomass and Bioenergy 66, 410-418. doi:10.1016/j.biombioe.2014.04.014.

Marvin, W.A., Schmidt, L.D., Benjaafar, S., Tiffany, D.G., Daoutidis, P., 2012. Economic Optimization of a Lignocellulosic Biomass-to-Ethanol Supply Chain. Chemical Engineering Science 67, 68-79. doi:10.1016/j.ces.2011.05.055.

Matas Güell, B., Bugge, M., Kempegowda, R. S., George, A., and Paap, S. M., 2012. Benchmark of Conversion and Production Technologies for Synthetic Biofuels for Aviation . Avinor project. Sintef, Trondheim, Norway.

Mathews, J.A., Tan, H., Moore, M.J.B., Bell, G., 2011. A conceptual lignocellulosic "feed fuel" biorefinery and its application to the linked biofuel and cattle raising industries in Brazil. Energy Policy 39, 4932-4938. doi:10.1016/j.enpol.2011.06.022.

McGrath, J., 2013a. Mallee jet fuel moves closer to reality. Future Farm 2013, 16-17.

McGrath, J., 2013b. Biofuel project moves beyond mallees. Future Farm 2013, 3.

McKendry, P., 2002a. Energy production from biomass (part 1): overview of biomass. Bioresource Technology. doi:10.1016/S0960-8524(01)00118-3.

McKendry, P., 2002b. Energy production from biomass (part 2): conversion technologies. Bioresource Technology. doi:10.1016/S0960-8524(01)00119-5.

McKendry, P., 2002c. Energy production from biomass (part 3): gasification technologies. Bioresource Technology. doi:10.1016/S0960-8524(01)00120-1.

Mello, F., Cerri, C., Davies, C.A., Holbrook, N.M., 2014. Payback time for soil carbon and sugar-cane ethanol. Nature Climate Change 4, 605-609. doi:10.1038/nclimate2239.

Meyer, P.M., Rodrigues, P.H.M., Rodrigues, P., Millen, D.D., 2013. Impact of biofuel production in Brazil on the economy, agriculture, and the environment. Animal Frontiers 3, 28-37. doi: 10.2527/af.2013-0012.

Mielenz, J.R., 2009. Biofuels - Methods and Protocols. Humana Press.

Milbrandt, A., McCormick, R.L., Kinchin, C.M., 2013. The Feasibility of Producing and Using Biomass-based Diesel and Jet Fuel in the United States (No. NREL/TP-6A20-58015). National Renewable Energy Laboratory, Golden, USA.

Morser, F., Soucacos, P., Hileman, J., Donohoo, P., Webb, S., 2011. Handbook for Analyzing the Costs and Benefits of Alternative Aviation Turbine Engine Fuels at Airports. National Academies Press, Washington DC, USA.

Mousdale, D.M., 2008. Biofuels: Biotechnology, Chemistry, and Sustainable Development. CRC Press, Boca Raton, USA.

Müller-Langer, F., Majer, S., O'Keeffe, S., 2014. Benchmarking biofuels—a comparison of technical, economic and environmental indicators. Energy, Sustainability and Society 4, 523. doi:10.1186/s13705-014-0020-x.

Nygren, E., Aleklett, K., Höök, M., 2009. Aviation fuel and future oil production scenarios. Energy Policy 37, 4003-4010. doi:10.1016/j.enpol.2009.04.048.

Parente, E., 2007. Lipofuels: biodiesel and biokerosene, Presented at the Biofuels from vegetal and animal oil as methyl or ethyl esthers, Tecbio, Fortaleza, Brazil, pp. 1-25.

Paterson, J., 2011. Mightier mallees make steady progress. Future Farm 2011, 10-11.

Perrin, R., Sesmero, J., Wamisho, K., Bacha, D., 2012. Biomass supply schedules for Great Plains delivery points. Biomass and Bioenergy 37, 213-220. doi:10.1016/j.biombioe. 2011.12.010.

Porras, F., 2008. Facts about bio-fueling in commercial aviation. Aviat Space Environ Med 79, 1078-1078.

Pray, T., 2010. Biomass research and development technical advisory committee: drop-in fuels panel, Presented at the Amyris meetings, Amyris. Department of Energy Report, Aurora, USA, pp. 1-20.

Rahmes, T., 2004. Status of Sustainable Biofuel Efforts for Aviation, Presented at the Boeing meetings, University of Washington, pp. 1-20.

Ranta, T., Korpinen, O.-J., Jäppinen, E., Karttunen, K., 2012. Forest Biomass Availability Analysis and Large-Scale Supply Options. Open Journal of Forestry 02, 33-40. doi:10.4236/ojf.2012.21005

Rentizelas, A.A., Tolis, A.J., Tatsiopoulos, I.P., 2009. Logistics issues of biomass: The storage problem and the multi-biomass supply chain. Renewable and Sustainable Energy Reviews 13, 887-894. doi:10.1016/j.rser.2008.01.003.

Richard, T.L., 2010. Challenges in scaling up biofuels infrastructure. Science. doi:10.1126/science.1189139.

Ringbeck, J., Koch, V., 2010. Aviation biofuels: a roadmap towards more carbon-neutral skies. Biofuels 1, 519-521. doi:10.4155/bfs.10.35.

Roni, M., Eksioglu, S., Cafferty, K.G., 2014. A Multi-Objective, Hub-and-Spoke Supply Chain Design Model For Densified Biomass (No. INL/CON-14-31477). Idaho National Laboratory, Idaho Falls, USA.

Rosillo-Calle, F., Teelucksingh, S., Thrän, D., Seiffert, M., 2012. The Potential Role of Biofuels in Commercial Air Transport - Biojetfuel. Imperial College, London, UK.

Rostrup-Nielsen, J.R., 2005. Making Fuels from Biomass. Science 308, 1421-1422. doi:10.1126/science.1113354.

Ruth, M., Mai, T., Newes, E., Aden, A., Warner, E., Uriarte, C., Inman, D., Simpkins, T., Argo, A., 2013. Projected biomass utilization for fuels and power in a mature market. National Renewable Energy Laboratory, Golden, USA.

Rye, L., Blakey, S., Wilson, C.W., 2010. Sustainability of supply or the planet: a review of potential drop-in alternative aviation fuels. Energy Environ. Sci. 3, 17-27. doi:10.1039/b918197k.

Sanke, N., Reddy, D.N., 2008. Biomass for power and energy generation. ICREPQ.

Schulze, E.D., Körner, C., Law, B.E., Haberl, H., Luyssaert, S., 2012. Large-scale bioenergy from additional harvest of forest biomass is neither sustainable nor greenhouse gas neutral. GCB Bioenergy 4, 611-616. doi:10.1111/j.1757-1707.2012.01169.x.

Sgouridis, S., 2012. Are we on course for a sustainable biofuel-based aviation future? Biofuels 3, 243-246. doi:10.4155/bfs.12.18.

Sgouridis, S., Bonnefoy, P.A., Hansman, R.J., 2011. Air transportation in a carbon constrained world: Long-term dynamics of policies and strategies for mitigating the carbon footprint of commercial aviation. Transportation Research Part A: Policy and Practice 45, 1077-1091. doi:10.1016/j.tra.2010.03.019.

Sims, R., Mabee, W., Saddler, J.N., Taylor, M., 2010. An overview of second generation biofuel technologies. Bioresource Technology. doi:10.1016/j.biortech.2009.11.046

Smith, A., Granda, C., Holtzapple, M., 2008. Biofuels for transportation, in: Kutz, M. (Ed.), Environmentally Conscious Transportation. John Wiley & Sons, pp. 213-256.

Smith, P., 2009. Biodiversity benefits of oil mallees.

Spatari, S., Bagley, D.M., MacLean, H.L., 2010. Life cycle evaluation of emerging lignocellulosic ethanol conversion technologies. Bioresource Technology 101, 654-667. doi:10.1016/j.biortech.2009.08.067.

Stanley, I., Bartle, J., 2013. Seeing the WA wheatbelt's future in trees. Future Farm 2013, 20-21.

Stucley, C., Schuck, S., Sims, R., Bland, J., Marino, B., Borowitzka, M., Abadi, A., Bartle, J., Giles, R., Thomas, Q., 2012. Bioenergy in Australia: status and opportunities. Bioenergy Australia (Forum) Limited, Surrey Hills, Australia.

Sultana, A., Kumar, A., 2014. Development of tortuosity factor for assessment of lignocellulosic biomass delivery cost to a biorefinery. Applied Energy 119, 288-295. doi:10.1016/j.apenergy.2013.12.036

Tao, L., Aden, A., 2009. The economics of current and future biofuels. In Vitro Cellular & Developmental Biology - Plant 45, 199-217. doi:10.1007/s11627-009-9216-8.

Taylor, G., 2008. Biofuels and the biorefinery concept. Energy Policy 36, 4406-4409. doi:10.1016/j.enpol.2008.09.069.

Turhollow, A., Downing, M., Butler, J., 1998. Forage Harvest and Transport Costs (No. ORNL/TM-13724). Oak Ridge National Laboratory, Springfields, USA.

Turnbull, P., 2012. Getting the best from mallee belts. Future Farm 2012, 4-5.

Veringa, H.J., Alderliesten, P., 2004. Advanced techniques for generation of energy from biomass and waste. Energy Research Center of the Netherlands (ECN), Petten, Netherlands.

Vimmerstedt, L.J., Bush, B.W., 2013. Effects of Deployment Investment on the Growth of the Biofuels Industry (No. NREL/TP-6A20-60802). National Renewable Energy Laboratory, Golden, USA. doi:10.2172/1118095.

Wang, C., Yoon, S.H., Jang, H.J., Chung, Y.R., Kim, J.Y., 2011. Metabolic engineering of Escherichia coli for α-farnesene production. Metabolic Engineering 13, 648-655. doi: 10.1016/j.ymben.2011.08.001.

Warshay, B., Pan, J., Sgouridis, S., 2010. Aviation industrys quest for a sustainable fuel: considerations of scale and modal opportunity carbon benefit. Biofuels 2, 33-58. doi: 10.4155/bfs.10.70.

Weber, G.F., Zygarlicke, C.J., 2001. Barrier issues to the utilization of biomass. University of North Dakota, Pittsburgh, USA.

Werner, D., 2014. Biofuels now. Aerospace America 30-35.

Wright, L.L., Boundy, R.G., Badger, P.C., Perlack, R.D., Davis, S.C., 2009. Biomass Energy Data Book (No. ORNL/TM-2009). Oak Ridge National Laboratory, Oak Ridge.

Yamada, Y., Inoue, G., Ono, M., 2013. Study of International Air-Passenger Traffic in East and Southeast Asia (No. 744). National Institute for Land and Infrastructure Management, Yokosuka, Japan.

Index

C

D

E

F

G

H

I

J

K

L

M

N

O

P

Q

R

S

T

U

V

W

X

Y

Z

www.ingramcontent.com/pod-product-compliance
Ingram Content Group UK Ltd.
Pitfield, Milton Keynes, MK11 3LW, UK
UKHW021653190726
13853UKWH00001B/236

9 782876 147065